4

Cambridge Tracts in Mathematics and Mathematical Physics

GENERAL EDITORS
J. G. LEATHEM, M.A.
E. T. WHITTAKER, M.A., F.R.S.

I0475286

No. 12

ORDERS OF INFINITY

THE 'INFINITÄRCALCÜL' OF PAUL DU BOIS-REYMOND

by

G. H. HARDY, M.A., F.R.S.
Fellow and Lecturer of Trinity College, Cambridge

Cambridge University Press
C. F. CLAY, Manager
London : Fetter Lane, E.C.
Edinburgh : 100, Princes Street
1910

L. C

Price 2s. 6d. *net*

CAMBRIDGE UNIVERSITY PRESS
London: FETTER LANE, E.C.
C. F. CLAY, Manager

Edinburgh: 100, PRINCES STREET
Berlin: A. ASHER AND CO.
Leipzig: F. A. BROCKHAUS
New York: G. P. PUTNAM'S SONS
Bombay and Calcutta: MACMILLAN AND CO., Ltd.

ORDERS OF INFINITY

THE 'INFINITÄRCALCÜL' OF
PAUL DU BOIS-REYMOND

by

G. H. HARDY, M.A., F.R.S.
Fellow and Lecturer of Trinity College, Cambridge

Cambridge:
at the University Press

1910

Cambridge:

PRINTED BY JOHN CLAY, M.A.

AT THE UNIVERSITY PRESS

PREFACE

THE ideas of Du Bois-Reymond's *Infinitärcalcül* are of great and growing importance in all branches of the theory of functions. With the particular system of notation that he invented, it is, no doubt, quite possible to dispense; but it can hardly be denied that the notation is exceedingly useful, being clear, concise, and expressive in a very high degree. In any case Du Bois-Reymond was a mathematician of such power and originality that it would be a great pity if so much of his best work were allowed to be forgotten.

There is, in Du Bois-Reymond's original memoirs, a good deal that would not be accepted as conclusive by modern analysts. He is also at times exceedingly obscure; his work would beyond doubt have attracted much more attention had it not been for the somewhat repugnant garb in which he was unfortunately wont to clothe his most valuable ideas. I have therefore attempted, in the following pages, to bring the *Infinitärcalcül* up to date, stating explicitly and proving carefully a number of general theorems the truth of which Du Bois-Reymond seems to have tacitly assumed—I may instance in particular the theorem of III. § 2.

I have to thank Messrs J. E. Littlewood and G. N. Watson for their kindness in reading the proof-sheets, and Mr J. Jackson for the numerical results contained in Appendix III.

G. H. H.

TRINITY COLLEGE,
April, 1910.

CONTENTS

I.

INTRODUCTION.

1. THE notions of the 'order of greatness' or 'order of smallness' of a function $f(n)$ of a positive integral variable n, when n is 'large,' or of a function $f(x)$ of a continuous variable x, when x is 'large' or 'small' or 'nearly equal to a,' are of the greatest importance even in the most elementary stages of mathematical analysis*. The student soon learns that as x tends to infinity $(x \to \infty)$ then also $x^2 \to \infty$, and moreover that x^2 tends to infinity *more rapidly than* x, *i.e.* that the ratio x^2/x tends to infinity as well; and that x^3 tends to infinity more rapidly than x^2, and so on indefinitely: and it is not long before he begins to appreciate the idea of a 'scale of infinity' (x^n) formed by the functions $x, x^2, x^3, ..., x^n, ...$ This scale he may supplement and to some extent complete by the interpolation of fractional powers of x, and, when he is familiar with the elements of the theory of the logarithmic and exponential functions, of irrational powers: and so he obtains a scale (x^a), where a is any positive number, formed by all possible positive powers of x. He then learns that there are functions whose rates of increase cannot be measured by any of the functions of this scale: that $\log x$, for example, tends to infinity more slowly, and e^x more rapidly, than *any* power of x; and that $x/(\log x)$ tends to infinity more slowly than x, but more rapidly than any power of x less than the first.

As we proceed further in analysis, and come into contact with its most modern developments, such as the theory of Fourier's series, the theory of integral functions, or the theory of singular points of analytic functions, the importance of these ideas becomes greater and

---

* See, for instance, my *Course of pure mathematics*, pp. 168 *et seq.*, 183 *et seq.*, 344 *et seq.*, 350.

greater. It is the systematic study of them, the investigation of
general theorems concerning them and ready methods of handling
them, that is the subject of Paul du Bois-Reymond's *Infinitärcalcül*
or 'calculus of infinities.'

2. The notion of the 'order' or the 'rate of increase' of a function
is essentially a relative one. If we wish to say that 'the rate of
increase of $f(x)$ is so and so' all we can say is that it is greater than,
equal to, or less than that of some other function $\phi(x)$.

Let us suppose that f and ϕ are two functions of the continuous
variable x, defined for all values of x greater than a given value x_0.
Let us suppose further that f and ϕ are positive, continuous, and
steadily increasing functions which tend to infinity with x; and let us
consider the ratio f/ϕ. We must distinguish four cases:

(i) If $f/\phi \to \infty$ with x, we shall say that the rate of increase, or
simply the *increase*, of f is greater than that of ϕ, and shall write

$$f \succ \phi.$$

(ii) If $f/\phi \to 0$, we shall say that the increase of f is less than that
of ϕ, and write

$$f \prec \phi.$$

(iii) If f/ϕ remains, for all values of x however large, between two
fixed positive numbers δ, Δ, so that $0 < \delta < f/\phi < \Delta$, we shall say that
the increase of f is equal to that of ϕ, and write

$$f \asymp \phi.$$

It may happen, in this case, that f/ϕ actually tends to a definite
limit. If this is so, we shall write

$$f \asymp \phi.$$

Finally, if this limit is *unity*, we shall write

$$f \sim \phi.$$

When we can compare the increase of f with that of some standard
function ϕ by means of a relation of the type $f \asymp \phi$, we shall say that
ϕ *measures*, or simply *is*, the increase of f. Thus we shall say that
the increase of $2x^2 + x + 3$ is x^2.

It usually happens in applications that f/ϕ is monotonic (*i.e.*
steadily increasing or steadily decreasing) as well as f and ϕ themselves. It is clear that in this case f/ϕ must tend to infinity, or zero,
or to a positive limit: so that one of the three cases indicated above

must occur, and we must have $f \succ \phi$ or $f \prec \phi$ or $f \asymp \phi$ (not merely $f \asymp \phi$). We shall see in a moment that this is not true in general.

(iv) It may happen that f/ϕ neither tends to infinity nor to zero, nor remains between fixed positive limits.

Suppose, for example, that ϕ_1, ϕ_2 are two continuous and increasing functions such that $\phi_1 \succ \phi_2$. A glance at the figure (Fig. 1) will probably show with sufficient clearness how we can construct, by means of a 'staircase' of straight or curved lines, running backwards and forwards between the graphs of ϕ_1 and ϕ_2, the graph of a steadily increasing function f such that $f = \phi_1$ for $x = x_1$, x_3, ... and $f = \phi_2$ for $x = x_2$, x_4, ... Then $f/\phi_1 = 1$ for

$$x = x_1, x_3, ...,$$

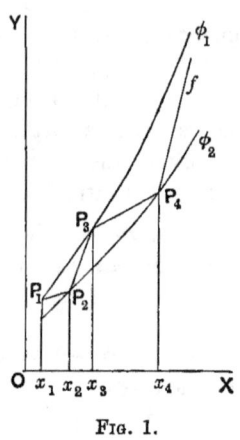

FIG. 1.

but assumes for $x = x_2$, x_4, ... values which decrease beyond all limit; while $f/\phi_2 = 1$ for $x = x_2$, x_4, ..., but assumes for $x = x_1$, x_3, ... values which increase beyond all limit; and f/ϕ, where ϕ is a function such that $\phi_1 \succ \phi \succ \phi_2$, as $e.g.$ $\phi = \sqrt{(\phi_1 \phi_2)}$, assumes both values which increase beyond all limit and values which decrease beyond all limit.

Later on (v. § 3) we shall meet with cases of this kind in which the functions are defined by explicit analytical formulae.

3. If a positive constant δ can be found such that $f > \delta\phi$ for all sufficiently large values of x, we shall write

$$f \succcurlyeq \phi ;$$

and if a positive constant Δ can be found such that $f < \Delta\phi$ for all sufficiently large values of x, we shall write

$$f \preccurlyeq \phi.$$

If $f \succcurlyeq \phi$ and $f \preccurlyeq \phi$, then $f \asymp \phi$.

It is however important to observe (i) that $f \succcurlyeq \phi$ is not logically equivalent to the negation of $f \prec \phi$* and (ii) that it is not logically equivalent to the alternative '$f \succ \phi$ or $f \asymp \phi$.' Thus, in the example discussed at the end of § 2, $\phi_1 \succcurlyeq f \succcurlyeq \phi_2$, but no one of the relations $\phi_1 \succ f$, etc. holds. If however we know that one of the relations $f \succ \phi$, $f \asymp \phi$, $f \prec \phi$ *must* hold, then these various assertions *are* logically equivalent.

* The relations $f \succcurlyeq \phi$, $f \prec \phi$ are mutually exclusive but not exhaustive: $f \succcurlyeq \phi$ implies the negation of $f \prec \phi$, but the converse is not true.

The reader will be able to prove without difficulty that the symbols $\succ$, $\asymp$, $\prec$ satisfy the following theorems.

$$\text{If } f \succ \phi,\; \phi \succcurlyeq \psi,\; \text{then } f \succ \psi.$$

$$\text{If } f \succcurlyeq \phi,\; \phi \succ \psi,\; \text{then } f \succ \psi.$$

$$\text{If } f \succcurlyeq \phi,\; \phi \succcurlyeq \psi,\; \text{then } f \succcurlyeq \psi.$$

$$\text{If } f \asymp \phi,\; \phi \asymp \psi,\; \text{then } f \asymp \psi.$$

$$\text{If } f \succcurlyeq \phi,\; \text{then } f + \phi \asymp f.$$

$$\text{If } f \succ \phi,\; \text{then } f - \phi \asymp f.$$

$$\text{If } f \succ \phi,\; f_1 \succ \phi_1,\; \text{then } f + f_1 \succ \phi + \phi_1.$$

$$\text{If } f \succ \phi,\; f_1 \asymp \phi_1,\; \text{then } f + f_1 \succcurlyeq \phi + \phi_1.$$

$$\text{If } f \asymp \phi,\; f_1 \asymp \phi_1,\; \text{then } f + f_1 \asymp \phi + \phi_1.$$

$$\text{If } f \succ \phi,\; f_1 \succcurlyeq \phi_1,\; \text{then } f f_1 \succ \phi \phi_1.$$

$$\text{If } f \asymp \phi,\; f_1 \asymp \phi_1,\; \text{then } f f_1 \asymp \phi \phi_1.$$

Many other obvious results of the same character might be stated, but these seem the most important. The reader will find it instructive to state for himself a series of similar theorems involving also the symbols $\asymp$ and $\sim$.

4. So far we have supposed that the functions considered all tend to infinity with x. There is nothing to prevent us from including also the case in which f or ϕ tends steadily to zero, or to a limit other than zero. Thus we may write $x \succ 1$, or $x \succ 1/x$, or $1/x \succ 1/x^2$. Bearing this in mind the reader should frame a series of theorems similar to those of § 3 but having reference to *quotients* instead of to sums or products.

It is also convenient to extend our definitions so as to apply to *negative* functions which tend steadily to $-\infty$ or to 0 or to some other limit. In such cases we make no distinction, when using the symbols $\succ$, $\prec$, $\asymp$, $\asymp$, between the function and its modulus: thus we write $-x \prec -x^2$ or $-1/x \prec 1$, meaning thereby exactly the same as by $x \prec x^2$ or $1/x \prec 1$. But $f \sim \phi$ is of course to be interpreted as a statement about the actual functions and not about their moduli.

It will be well to state at this point, once for all, that all functions referred to in this tract, from here onwards, are to be understood, unless the contrary is expressly stated or obviously implied, to be positive, continuous, and monotonic, increasing of course if they tend to ∞, and decreasing if they tend to 0. But it is sometimes con-

venient to use our symbols even when this is not true of all the
functions concerned; to write, for example,

$$1 + \sin x \prec x, \quad x^2 \succ x \sin x,$$

meaning by the first formula simply that $|1 + \sin x|/x \to 0$. This
kind of use may clearly be extended even to complex functions
(*e.g.* $e^{ix} \prec x$).

Again, we have so far confined our attention to functions of a
continuous variable x which tends to $+\infty$. This case includes that
which is perhaps even more important in applications, that of functions
of the positive integral variable n: we have only to disregard values of
x other than integral values. Thus $n! \succ n^2$, $-1/n \prec n$.

Finally, by putting $x = -y$, $x = 1/y$, or $x = 1/(y-a)$, we are led to
consider functions of a continuous variable y which tends to $-\infty$ or 0
or a: the reader will find no difficulty in extending the considerations
which precede to cases such as these.

In what follows we shall generally state and prove our theorems
only for the case with which we started, that of indefinitely increasing
functions of an indefinitely increasing continuous variable, and shall
leave to the reader the task of formulating the corresponding theorems
for the other cases. We shall in fact always adopt this course, except
on the rare occasions when there is some essential difference between
different cases.

5. There are some other symbols which we shall sometimes find it
convenient to use in special senses.

By $$O(\phi)$$

we shall denote a function f, otherwise unspecified, but such that

$$|f| < K\phi,$$

where K is a positive constant, and ϕ a positive function of x: this
notation is due to Landau. Thus

$$x + 1 = O(x), \quad x = O(x^2), \quad \sin x = O(1).$$

We shall follow Borel in using the same letter K in a whole series
of inequalities to denote a positive constant, not necessarily the same
in all inequalities where it occurs. Thus

$$\sin x < K, \quad 2x + 1 < Kx, \quad x^m < Ke^x.$$

If we use K thus in any finite number of inequalities which (like the
first two above) do not involve any variables other than x, or whatever
other variable we are primarily considering, then all the values of K lie

between certain absolutely fixed limits K_1 and K_2 (thus K_1 might be 10^{-10} and K_2 be 10^{10}). In this case all the K's satisfy $0 < K_1 < K < K_2$, and every relation $f < K\phi$ might be replaced by $f < K_2\phi$, and every relation $f > K\phi$ by $f > K_1\phi$. But we shall also have occasion to use K in equalities which (like the third above) involve a parameter (here m). In this case K, though independent of x, is a function of m. Suppose that a, β, ... are all the parameters which occur in this way in this tract. Then if we give any special system of values to a, β, ..., we can determine K_1, K_2 as above. Thus all our K's satisfy

$$0 < K_1(a, \beta, ...) < K < K_2(a, \beta, ...),$$

where K_1, K_2 are positive functions of a, β, ... defined for any permissible set of values of those parameters. But K_1 has zero for its lower limit; by choosing a, β, ... appropriately we can make K_1 as small as we please—and, of course, K_2 as large as we please*.

It is clear that the three assertions

$$f = O(\phi), \quad |f| < K\phi, \quad f \preccurlyeq \phi$$

are precisely equivalent to one another.

When a function f possesses any property for all values of x greater than some definite value (this value of course depending on the nature of the particular property) we shall say that f possesses the property for $x > x_0$. Thus

$$x > 100 \quad (x > x_0), \quad e^x > 100\, x^2 \quad (x > x_0).$$

We shall use δ to denote an arbitrarily small but fixed positive number, and Δ to denote an arbitrarily great but likewise fixed positive number. Thus

$$f < \delta\phi \quad (x > x_0)$$

means 'however small δ, we can find x_0 so that $f < \delta\phi$ for $x > x_0$,' i.e. means the same as $f \prec \phi$; and $\phi > \Delta f$ $(x > x_0)$ means the same: and

$$(\log x)^\Delta \prec x^\delta$$

means 'any power of $\log x$, however great, tends to infinity more slowly than any positive power of x, however small.'

Finally, we denote by ϵ a function (of a variable or variables indicated by the context or by a suffix) whose limit is zero when the variable or variables are made to tend to infinity or to their limits in the way we happen to be considering. Thus

$$f = \phi(1 + \epsilon), \quad f \sim \phi$$

are equivalent to one another.

* I am indebted to Mr Littlewood for the substance of these remarks.

In order to become familiar with the use of the symbols defined in the preceding sections the reader is advised to verify the following relations; in them $P_m(x)$, $Q_n(x)$ denote polynomials whose degrees are m and n and whose leading coefficients are positive:

$$P_m(x) \succ Q_n(x) \ (m > n), \quad P_m(x) \asymp Q_n(x) \ (m = n),$$

$$P_m(x) \asymp x^m, \quad P_m(x)/Q_n(x) \asymp x^{m-n},$$

$$\sqrt{(ax^2 + 2bx + c)} \asymp x \ (a > 0), \quad \sqrt{(x+a)} \sim \sqrt{x},$$

$$\sqrt{(x+a)} - \sqrt{x} \sim a/2\sqrt{x}, \quad \sqrt{(x+a)} - \sqrt{(x)} = O\,(1/\sqrt{x}),$$

$$e^x \succ x^\Delta, \quad e^{x^2} \succ e^{\Delta x}, \quad e^{e^x} \succ e^{x^\Delta},$$

$$\log x \prec x^\delta, \quad \log P_m(x) \asymp \log Q_n(x), \quad \log\log P_m(x) \sim \log\log Q_n(x),$$

$$x + a\sin x \sim x, \quad x\,(a + \sin x) \asymp x \ (a > 1),$$

$$e^{a + \sin x} \asymp 1, \quad \cosh x \sim \sinh x \asymp e^x,$$

$$x^m = O\,(e^{\delta x}), \quad (\log x)/x = O\,(x^{\delta - 1}),$$

$$1 + \frac{1}{2} + \dots + \frac{1}{n} \succ 1, \quad 1 + \frac{1}{2^2} + \dots + \frac{1}{n^2} \asymp 1,$$

$$1 + \frac{1}{2} + \dots + \frac{1}{n} \sim \log n, \quad 1 + \frac{1}{2} + \dots + \frac{1}{n} - \log n \asymp 1,$$

$$n! \prec n^n, \quad n! \succ e^{\Delta n}, \quad n! = n^{n^{1+\epsilon}} = n^{n(1+\epsilon)},$$

$$n! \sim n^{n+\frac{1}{2}} e^{-n} \sqrt{(2\pi)}, \quad n!\,(e/n)^n = (1+\epsilon)\sqrt{(2\pi n)},$$

$$\int_1^x \frac{dt}{t} \succ 1, \quad \int_1^x \frac{dt}{t} \sim \log x, \quad \int_2^x \frac{dt}{\log t} \sim \frac{x}{\log x}.$$

II.

SCALES OF INFINITY IN GENERAL.

1. If we start from a function ϕ, such that $\phi \succ 1$, we can, in a variety of ways, form a series of functions

$$\phi_1 = \phi, \quad \phi_2, \quad \phi_3, \dots, \quad \phi_n, \dots$$

such that the increase of each function is greater than that of its predecessor. Such a sequence of functions we shall denote for shortness by (ϕ_n).

One obvious method is to take $\phi_n = \phi^n$. Another is as follows: If $\phi \succ x$, it is clear that

$$\phi\{\phi(x)\}/\phi(x) \to \infty,$$

and so $\phi_2(x) = \phi\phi(x) \succ \phi(x)$; similarly $\phi_3(x) = \phi\phi_2(x) \succ \phi_2(x)$, and so on*.

Thus the first method, with $\phi = x$, gives the scale x, x^2, x^3, ... or (x^n); the second, with $\phi = x^2$, gives the scale x^2, x^4, x^8, ... or (x^{2^n}).

These scales are *enumerable* scales, formed by a simple progression of functions. We can also, of course, by replacing the integral parameter n by a continuous parameter a, define scales containing a non-enumerable multiplicity of functions : the simplest is (x^a), where a is any positive number. But such scales fill a subordinate *rôle* in the theory.

It is obvious that we can always insert a new term (and therefore, of course, any number of new terms) in a scale at the beginning or between any two terms : thus $\sqrt{\phi}$ (or ϕ^a, where a is any positive number less than unity) has an increase less than that of any term of the scale, and $\sqrt{(\phi_n\phi_{n+1})}$ or $\phi_n^a\phi_{n+1}^{1-a}$ has an increase intermediate between those of ϕ_n and ϕ_{n+1}. A less obvious and far more important theorem is the following

Theorem of Paul du Bois-Reymond. *Given any ascending scale of increasing functions ϕ_n, i.e. a series of functions such that $\phi_1 \prec \phi_2 \prec \phi_3 \prec$ ..., we can always find a function f which increases more rapidly than any function of the scale, i.e. which satisfies the relation $\phi_n \prec f$ for all values of n.*

In view of the fundamental importance of this theorem we shall give two entirely different proofs.

2. (i) We know that $\phi_{n+1} \succ \phi_n$ for all values of n, but this, of course, does not necessarily imply that $\phi_{n+1} \geqslant \phi_n$ for all values of x and n in question†. We can, however, construct a new scale of functions ψ_n such that

(a) ψ_n is identical with ϕ_n for all values of x from a certain value x_n onwards (x_n, of course, depending upon n);

(b) $\psi_{n+1} \geqslant \psi_n$ for all values of x and n.

For suppose that we have constructed such a scale up to its nth term ψ_n. Then it is easy to see how to construct ψ_{n+1}. Since $\phi_{n+1} \succ \phi_n$, $\phi_n \sim \psi_n$, it follows that $\phi_{n+1} \succ \psi_n$, and so $\phi_{n+1} > \psi_n$ from a certain value of x (say x_{n+1}) onwards. For $x \geqslant x_{n+1}$ we take $\psi_{n+1} = \phi_{n+1}$. For $x < x_{n+1}$ we give ψ_{n+1} a value equal to the greater of the values of

---

* For some results as to the increase of such iterated functions see VII. § 2 (vi).

† $\phi_{n+1} \succ \phi_n$ implies $\phi_{n+1} > \phi_n$ for sufficiently large values of x, say for $x > x_n$. But x_n may tend to ∞ with n. Thus if $\phi_n = x^n/n!$ we have $x_n = n+1$.

ϕ_{n+1}, ψ_n. Then it is obvious that ψ_{n+1} satisfies the conditions (a) and (b).

Now let $$f(n) = \psi_n(n).$$

From $f(n)$ we can deduce a continuous and increasing function $f(x)$, such that

$$\psi_n(x) < f(x) < \psi_{n+1}(x)$$

for $n < x < n+1$, by joining the points $(n, \psi_n(n))$ by straight lines or suitably chosen arcs of curves.

It is perhaps worth while to call attention explicitly to a small point that has sometimes been overlooked (see, e.g., Borel, *Leçons sur la théorie des fonctions*, p. 114; *Leçons sur les séries à termes positifs*, p. 26). It is not always the case that the use of straight lines will ensure

$$f(x) > \psi_n(x)$$

for $x > n$ (see, for example, Fig. 2, where the dotted line represents an appropriate arc).

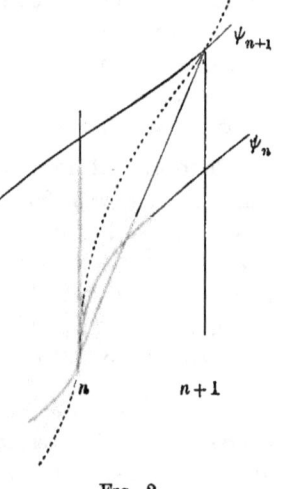

Then $$f/\psi_n > \psi_{n+1}/\psi_n$$

for $x > n+1$, and so $f \succ \psi_n$; therefore $f \succ \phi_n$, and the theorem is proved.

The proof which precedes may be made more general by taking $f(n) = \psi_{\lambda_n}(n)$, where λ_n is an integer depending upon n and tending steadily to infinity with n.

FIG. 2.

(ii) The second proof of Du Bois-Reymond's Theorem proceeds on entirely different lines. We can always choose positive coefficients a_n so that

$$f(x) = \sum_1^\infty a_n \psi_n(x)$$

is convergent for all values of x. This will certainly be the case, for instance, if

$$1/a_n = \psi_1(1)\psi_2(2) \dots \psi_n(n).$$

For then, if ν is any integer greater than x, $\psi_n(x) < \psi_n(n)$ for $n \geq \nu$, and the series will certainly be convergent if

$$\sum_\nu^\infty \frac{1}{\psi_1(1)\psi_2(2) \dots \psi_{n-1}(n-1)}$$

is convergent, as is obviously the case.

Also $f(x)/\psi_n(x) > a_{n+1}\psi_{n+1}(x)/\psi_n(x) \to \infty$,

so that $f \succ \phi_n$ for all values of n.

3. Suppose, *e.g.*, that $\phi_n = x^n$. If we restrict ourselves to values of x greater than 1, we may take $\psi_n = \phi_n = x^n$. The first method of construction would naturally lead to

$$f = n^n = e^{n \log n},$$

or $f = (\lambda_n)^n$, where λ_n is defined as at the end of § 2 (i), and each of these functions has an increase greater than that of any power of n. The second method gives

$$f(x) = \sum_1^\infty \frac{x^n}{1^1 2^2 3^3 \dots n^n}.$$

It is known* that when x is large the order of magnitude of this function is roughly the same as that of

$$e^{\frac{1}{2} (\log x)^2 / \log \log x}.$$

As a matter of fact it is by no means necessary, in general, in order to ensure the convergence of the series by which $f(x)$ is defined, to suppose that a_n decreases so rapidly. It is very generally sufficient to suppose $1/a_n = \phi_n(n)$: this is always the case, for example, if $\phi_n(x) = \{\phi(x)\}^n$, as the series

$$\sum \left\{\frac{\phi(x)}{\phi(n)}\right\}^n$$

is always convergent. This choice of a_n would, when $\phi = x$, lead to

$$f(x) = \sum \left(\frac{x}{n}\right)^n \sim \sqrt{\left(\frac{2\pi x}{e}\right)} e^{x/e} \dagger.$$

But the simplest choice here is $1/a_n = n!$, when

$$f(x) = \sum \frac{x^n}{n!} = e^x - 1 ;$$

it is naturally convenient to disregard the irrelevant term -1.

4. We can always suppose, if we please, that $f(x)$ is defined by a power series $\sum a_n x^n$ convergent for all values of x, in virtue of a theorem of Poincaré's ‡ which is of sufficient intrinsic interest to deserve a formal statement and proof.

Given any continuous increasing function $\phi(x)$, *we can always find an integral function* $f(x)$ *(i.e. a function* $f(x)$ *defined by a power series* $\sum a_n x^n$ *convergent for all values of* x*) such that* $f(x) \succ \phi(x)$.

The following simple proof is due to Borel §.

Let $\Phi(x)$ be any function (such as the square of ϕ) such that $\Phi \succ \phi$. Take

---

* *Messenger of Mathematics*, vol. 34, p. 101.

† Lindelöf, *Acta Societatis Fennicae*, t. 31, p. 41; Le Roy, *Bulletin des Sciences Mathématiques*, t. 24, p. 245.

‡ *American Journal of Mathematics*, vol. 14, p. 214.

§ *Leçons sur les séries à termes positifs*, p. 27.

an increasing sequence of numbers a_n such that $a_n \to \infty$, and another sequence of numbers b_n such that

$$a_1 < b_2 < a_2 < b_3 < a_3 < \ldots\ldots;$$

and let $$f(x) = \Sigma \left(\frac{x}{b_n}\right)^{\nu_n},$$

where ν_n is an integer and $\nu_{n+1} > \nu_n$. This series is convergent for all values of x; for the nth root of the nth term is, for sufficiently large values of n, not greater than x/b_n, and so tends to zero. Now suppose $a_n \leqq x < a_{n+1}$; then

$$f(x) > \left(\frac{a_n}{b_n}\right)^{\nu_n}.$$

Since $a_n > b_n$ we can suppose ν_n so chosen that (i) ν_n is greater than any of $\nu_1, \nu_2, \ldots, \nu_{n-1}$ and (ii)

$$\left(\frac{a_n}{b_n}\right)^{\nu_n} > \Phi(a_{n+1}).$$

Then $$f(x) > \Phi(a_{n+1}) > \Phi(x),$$

and so $f \succ \phi$.

5. So far we have confined our attention to ascending scales, such as $x, x^2, x^3, \ldots, x^n, \ldots$ or (x^n); but it is obvious that we may consider in a similar manner *descending* scales such as $x, \sqrt{x}, \sqrt[3]{x}, \ldots, \sqrt[n]{x}, \ldots$ or $(\sqrt[n]{x})$. It is very generally (though not always) true that if (ϕ_n) is an ascending scale, and ψ denotes the function inverse to ϕ, then (ψ_n) is a descending scale.

If $\phi > \bar{\phi}$ for all values of x (or all values greater than some definite value), then a glance at Fig. 3 is enough to show that if ψ and $\bar{\psi}$ are the functions inverse to ϕ and $\bar{\phi}$, then $\psi < \bar{\psi}$ for all values of x (or all values greater than some definite value). We have only to remember that the graph of ψ may be obtained from that of ϕ by looking at the latter from a different point of view (interchanging the *rôles* of x and y). But it is not true that $\phi \succ \bar{\phi}$ involves $\psi \prec \bar{\psi}$. Thus $e^x \succ e^x/x$. The function inverse to e^x is $\log x$: the function inverse to e^x/x is obtained by solving the equation $x = e^y/y$ with respect to y. This equation gives

$$y = \log x + \log y,$$

and it is easy to see that $y \sim \log x$.

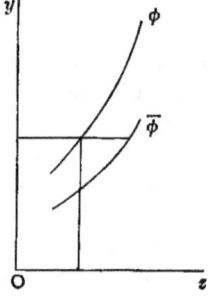

FIG. 3.

Given a scale of increasing functions ϕ_n such that

$$\phi_1 \succ \phi_2 \succ \phi_3 \succ \ldots \succ 1,$$

we can find an increasing function f such that $\phi_n \succ f \succ 1$ *for all values of n.* The reader will find no difficulty in modifying the argument of § 2 (i) so as to establish this proposition.

6. The following extensions of Du Bois-Reymond's Theorem (and the corresponding theorem for descending scales) are due to Hadamard [*].

Given $\qquad \phi_1 \prec \phi_2 \prec \phi_3 \prec \ldots \prec \phi_n \prec \ldots \prec \Phi,$

we can find f so that $\phi_n \prec f \prec \Phi$ *for all values of n.*

Given $\qquad \psi_1 \succ \psi_2 \succ \psi_3 \succ \ldots \succ \psi_n \succ \ldots \succ \Psi,$

we can find f so that $\psi_n \succ f \succ \Psi$ *for all values of n.*

Given an ascending sequence (ϕ_n) *and a descending sequence* (ψ_p) *such that* $\phi_n \prec \psi_p$ *for all values of n and p, we can find f so that*

$$\phi_n \prec f \prec \psi_p$$

for all values of n and p.

To prove the first of these theorems we have only to observe that

$$\Phi/\phi_1 \succ \Phi/\phi_2 \succ \ldots \succ \Phi/\phi_n \succ \ldots \succ 1,$$

and to construct a function F (as we can in virtue of the theorem of § 5) which tends to infinity more slowly than any of the functions Φ/ϕ_n. Then

$$f = \Phi/F$$

is a function such as is required. Similarly for the second theorem. The third is rather more difficult to prove.

In the first place, we may suppose that $\phi_{n+1} > \phi_n$ for all values of x and n: for if this is not so we can modify the definitions of the functions ϕ_n as in § 2 (i). Similarly we may suppose $\psi_{p+1} < \psi_p$ for all values of x and p.

Secondly, we may suppose that, if x is fixed, $\phi_n \to \infty$ as $n \to \infty$, and $\psi_p \to 0$ as $p \to \infty$. For if this is not true of the functions given, we can replace them by $H_n\phi_n$ and $K_p\psi_p$, where (H_n) is an increasing sequence of constants, tending to ∞ with n, and (K_p) a decreasing sequence of constants whose limit as $p \to \infty$ is zero.

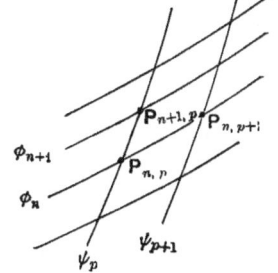

FIG. 4.

Since $\psi_p \succ \phi_n$ but, for any given x, $\psi_p < \phi_n$ for sufficiently large values of n, it is clear (see Fig. 4) that the curve $y = \psi_p$ intersects the curve $y = \phi_n$ for all sufficiently large values of n (say for $n \geqslant n_p$).

* *Acta Mathematica*, t. 18, pp. 319 *et seq.*

At this point we shall, in order to avoid unessential detail, introduce a restrictive hypothesis which can be avoided by a slight modification of the argument*, but which does not seriously impair the generality of the result. We shall assume that no curve $y = \psi_p$ intersects any curve $y = \phi_n$ in more than one point; let us denote this point, if it exists, by $P_{n,p}$.

If p is fixed, $P_{n,p}$ exists for $n > n_p$; similarly, if n is fixed, $P_{n,p}$ exists for $p > p_n$. And as either n or p increases, so do both the ordinate or the abscissa of $P_{n,p}$. The curve ψ_p contains all the points $P_{n,p}$ for which p has a fixed value: and $y = \phi_n$ contains all the points for which n has a fixed value.

It is clear that, in order to define a function f which tends to infinity more rapidly than any ϕ_n and less rapidly than any ψ_p, all that we have to do is to draw a curve, making everywhere a positive acute angle with each of the axes of coordinates, and crossing all the curves $y = \phi_n$ from below to above, and all the curves $y = \psi_p$ from above to below.

Choose a positive integer N_p, corresponding to each value of p, such that (i) $N_p > n_p$ and (ii) $N_p \to \infty$ as $p \to \infty$. Then $P_{N_p, p}$ exists for each value of p. And it is clear that we have only to join the points $P_{N_1,1}, P_{N_2,2}, P_{N_3,3}, \ldots$ by straight lines or other suitably chosen arcs of curves in order to obtain a curve which fulfils our purpose. The theorem is therefore established.

7. Some very interesting considerations relating to scales of infinity have been developed by Pincherle[†].

We have defined $f \succ \phi$ to mean $f/\phi \to \infty$, or, what is the same thing,

$$\log f - \log \phi \to \infty \qquad \ldots\ldots\ldots\ldots\ldots\ldots(1).$$

We might equally well have defined $f \succ \phi$ to mean

$$F(f) - F(\phi) \to \infty \qquad \ldots\ldots\ldots\ldots\ldots\ldots(2),$$

where $F(x)$ is any function which tends steadily to infinity with x (e.g. x, e^x). Let us say that if (2) holds then

$$f \succ \phi \ (F) \ldots\ldots\ldots\ldots\ldots\ldots\ldots(3),$$

so that $f \succ \phi$ is equivalent to $f \succ \phi$ (log x). Similarly we define $f \prec \phi$ (F) to mean that $F(f) - F(\phi) \to -\infty$, and $f \asymp \phi$ (F) to mean that $F(f) - F(\phi)$ remains between certain fixed limits. Thus

$$x + \log x \asymp x, \quad x + \log x \succ x \ (x),$$

$$x + 1 \asymp x \ (x), \quad x + 1 \succ x \ (e^x),$$

since $e^{x+1} - e^x = (e-1)e^x \to \infty$.

---

* See Hadamard's original paper quoted above.

† *Memorie della Accademia delle Scienze di Bologna* (ser. 4, t. 5, p. 739).

It is clear that the more rapid the increase of F, the more likely is it to discriminate between the rates of increase of two given functions f and ϕ. More precisely, *if*

$$f \succ \phi \ (F),$$

and $\overline{F} = FF_1$, *where F_1 is any increasing function, then will*

$$f \succ \phi \ (\overline{F}).$$

For

$$\overline{F}(f) - \overline{F}(\phi) = F(f) F_1(f) - F(\phi) F_1(\phi) > \{F(f) - F(\phi)\} F_1(\phi) \to \infty.$$

8. The substance of the following theorems is due in part to Pincherle and in part to Du Bois-Reymond*.

1. *However rapid the increase of f, as compared with that of ϕ, we can so choose F that $f \asymp \phi \ (F)$.*

2. *If $f - \phi$ is positive for $x > x_0$, we can so choose F that $f \succ \phi \ (F)$.*

3. *If $f - \phi$ is monotonic and not negative for $x > x_0$, and $f \asymp \phi \ (F)$, however great be the increase of F, then $f = \phi$ from a certain value of x onwards.*

(1) If $f \succ \phi$, we may regard f as an increasing function of ϕ, say

$$f = \theta(\phi),$$

where $\theta(x) \succ x$. We can choose a constant g greater than 1, and then choose X so that $\theta(x) > gx$ for $x > X$. Let a be any number greater than X, and let

$$a_1 = \theta(a), \ a_2 = \theta(a_1), \ a_3 = \theta(a_2), \dots.$$

Then (a_n) is an increasing sequence, and $a_n \to \infty$, since $a_n > g^n a$. We can now construct an increasing function F such that

$$F(a_n) = \tfrac{1}{2} n K,$$

where K is a constant. Then if $a_{\nu-1} \leqq x \leqq a_\nu, \ a_\nu \leqq \theta(x) \leqq a_{\nu+1}$, and

$$F\{\theta(x)\} - F(x) < F(a_{\nu+1}) - F(a_{\nu-1}) < K.$$

Accordingly $F(f) - F(\phi)$ remains less than a constant, and so the first theorem is established.

(2) Let $f - \phi = \lambda$, so that $\lambda > 0$. If λ, as x increases, remains greater than a constant K, then

$$e^f - e^\phi > (e^K - 1) e^\phi \to \infty,$$

so that we may take $F(x) = e^x$.

---

* Pincherle, *l.c.*; Du Bois-Reymond, *Math. Annalen*, Bd. 8, S. 390 *et seq.*

If it is not true that $\lambda \geq K$, λ assumes values less than any assignable positive number, as $x \to \infty$. Let $\bar{\lambda}(x)$ be defined as the lower limit of $\lambda(\xi)$ for $\xi \leq x$. Then $\bar{\lambda}$ tends steadily to zero as $x \to \infty$, and $\bar{\lambda} \leq \lambda$. We may also regard $\bar{\lambda}$ as a steadily decreasing function of ϕ, say $\bar{\lambda} = \mu(\phi)$.

Let $\varpi(\phi)$ be an increasing function of ϕ such that $\varpi \succ 1/\mu$, $\mu\varpi \succ 1$. Then if

$$F = \int^{\phi} \varpi(t)\,dt,$$

$$F(f) - F(\phi) = \int_{\phi}^{\phi+\lambda} \varpi\,dt \geq \int_{\phi}^{\phi+\mu(\phi)} \varpi\,dt > \mu(\phi)\varpi(\phi) \succ 1,$$

and $F(x)$ fulfils the requirement of theorem 2. The third theorem is obviously a mere corollary of the second.

The reader will find it instructive to deduce or prove independently the following three theorems, which are closely analogous to those which have just been proved.

1. *However great be the increase of f as compared with that of ϕ, we can determine an increasing function F such that $F(f) \asymp F(\phi)$.*

2. *If $f - \phi$ is positive for $x > x_0$, we can determine an increasing function F such that $F(f) \succ F(\phi)$.*

3. *If $f - \phi$ is monotonic and not negative for $x > x_0$, and $F(f) \asymp F(\phi)$, however great the increase of F, then $f = \phi$ from a certain value of x onwards.*

To these he may add the theorem (analogous to that proved at the end of § 7) that $f \succ \phi$ *involves $F(f) \succ F(\phi)$ if $\log F(x)/\log x$ is an increasing function* (a condition which may for practical purposes be replaced by $F \succ x$).

9. Let us consider some examples of the theorems of the last paragraph.

(i) Let $f = x^m$ ($m > 1$) and $\phi = x$. Then, following the argument of § 8 (1), we have $\theta(\phi) = \phi^m$. We may take

$$a = e,\ a_1 = e^m,\ a_2 = e^{m^2},\ \dots,\ a_n = e^{m^n},\ \dots,$$

and we have to define F so that

$$F(e^{m^n}) = \tfrac{1}{2}nK.$$

The most natural solution of this equation is

$$F(x) = K \log \log x / 2 \log m.$$

And in fact

$$F(x^m) - F(x) = \frac{K}{2 \log m}\{\log(m \log x) - \log \log x\} = \tfrac{1}{2}K,$$

so that $x^m \asymp x$ (F).

(ii) Let $f = e^x + e^{-x}$, $\phi = e^x$. Following the argument of § 8 (2), we find

$$\lambda = e^{-x} = \bar{\lambda}, \quad \mu(\phi) = 1/\phi,$$

and we may take $\varpi(\phi) = \phi^{1+a}$ $(a > 0)$. This makes F a constant multiple of x^{2+a}, and it is easy to verify that

$$(e^x + e^{-x})^k - e^{kx} \to \infty,$$

if $k > 2$.

(iii) The relation $F(f) \asymp F(\phi)$ is equivalent to $f \asymp \phi$ $(\log F)$. Using the result of (i) we see that $F(x^m) \asymp F(x)$ if $F \prec \log x$. Similarly, using the result of (ii), we see that $F(e^x + e^{-x}) \succ F(e^x)$ if $F \succ e^{x^k}$ $(k > 2)$.

10. Before leaving this part of our subject, let us observe that all of the substance of §§ 1-6 of this section may be extended to the case in which our symbols $\succ$, etc., are defined by reference to an arbitrary increasing function F. We leave it as an exercise to the reader to effect these extensions.

III.

LOGARITHMICO-EXPONENTIAL SCALES.

1. The only scales of infinity that are of any practical importance in analysis are those which may be constructed by means of the logarithmic and exponential functions.

We have already seen (II. § 3) that

$$e^x \succ x^n$$

for any value of n however great. From this it follows that

$$\log x \prec x^{1/m}$$

for any value of n *.

It is easy to deduce that

$$e^{e^x} \succ e^{x^n}, \quad e^{e^{e^x}} \succ e^{e^{x^n}}, \dots,$$

$$\log \log x \prec (\log x)^{1/n}, \quad \log \log \log x \prec (\log \log x)^{1/n}, \dots.$$

---

* It was pointed out above (II. § 5) that $\phi \succ \bar{\phi}$ does not necessarily involve $\psi \prec \bar{\psi}$ (ψ, $\bar{\psi}$ being the functions inverse to ϕ, $\bar{\phi}$). But it does involve $\psi < \bar{\psi}$ for sufficiently large values of x, and therefore $\psi \leqslant \bar{\psi}$. Hence $\phi \succ \phi_n$ (for any n) involves $\psi \leqslant \psi_n$ (for any n) and therefore, if (ψ_n) is a descending scale, as is in this case obvious, $\psi \prec \psi_n$ for any n. For proofs of the relations $e^x \succ x^n$, $\log x \prec x^{1/n}$, proceeding on different lines, see my *Course of pure mathematics*, pp. 345, 350.

The repeated logarithmic and exponential functions are so important in this subject that it is worth while to adopt a notation for them of a less cumbrous character. We shall write

$$l_1 x \equiv lx \equiv \log x, \quad l_2 x \equiv llx, \quad l_3 x \equiv ll_2 x, \ldots,$$
$$e_1 x \equiv ex \equiv e^x, \quad e_2 x \equiv eex, \quad e_3 x \equiv ee_2 x, \ldots.$$

It is easy, with the aid of these functions, to write down any number of ascending scales, each containing only functions whose increase is greater than that of any function in any preceding scale; for example

$$x, x^2, \ldots, x^n, \ldots; \quad e^x, e^{2x}, \ldots, e^{nx}, \ldots;$$
$$e^{x^2}, e^{x^3}, \ldots, e^{x^n}, \ldots; \quad e_2 x, e_3 x, \ldots, e_n x, \ldots.$$

In among the functions of these scales we can of course interpolate new functions as freely as we like, using, for instance, such functions as

$$x^\alpha e^{\beta x^\gamma} e^{\delta x^\epsilon},$$

where α, β, γ, δ, ϵ are any positive numbers; and we can of course construct non-enumerable (II. § 1) as well as enumerable scales. Similarly we can construct any number of descending scales, each composed of functions whose increase is less than that of any functions in any preceding scale: for example

$$lx, (lx)^{1/2}, \ldots, (lx)^{1/n}, \ldots; \quad l_2 x, l_3 x, \ldots, l_n x, \ldots.$$

Two special scales are of particularly fundamental importance; the ascending scale

$$(E) \qquad x, \ ex, \ e_2 x, \ e_3 x, \ldots,$$

and the descending scale

$$(L) \qquad x, \ lx, \ l_2 x, \ l_3 x, \ldots.$$

These scales mark the *limits* of all logarithmic and exponential scales : it is of course, in virtue of the general theorems of II., possible to define functions whose increase is more rapid than that of any $e_n x$ or slower than that of any $l_n x$; but, as we shall see in a moment, this is *not* possible if we confine ourselves to functions defined by a finite and explicit formula involving only the ordinary functional symbols of elementary analysis.

2. We define a *logarithmico-exponential function* (shortly, an *L-function*) as a real one-valued function defined, for all values of x greater than some definite value, by a finite combination of the ordinary algebraical symbols (viz. $+$, $-$, $\times$, $\div$, $\sqrt[n]{\ }$) and the functional symbols $\log(\ldots)$ and $e^{(\cdots)}$, operating on the variable x and on real constants.

It is to be observed that the result of working out the value of the function, by substituting x in the formula defining it, is to be real at all stages of the work. It is important to exclude such a function

$$\tfrac{1}{2}\{e^{\sqrt{(-x^2)}} + e^{-\sqrt{(-x^2)}}\},$$

which, with a suitable interpretation of the roots, is equal to $\cos x$.

Theorem. *Any L-function is ultimately continuous, of constant sign, and monotonic, and, as $x \to \infty$, tends to ∞, or to zero or to some other definite limit. Further, if f and ϕ are L-functions, one or other of the relations*

$$f \succ \phi, \; f \asymp \phi, \; f \prec \phi$$

holds between them.

We may classify L-functions as follows, by a method due to Liouville*. An L-function is of order zero if it is purely algebraical; of order 1 if the functional symbols $l(\ldots)$ and $e(\ldots)$ which occur in it bear only on algebraical functions; of order 2 if they bear only on algebraical functions or L-functions of order 1; and so on. Thus

$$x^{x^x} = e^{\log x\, e^{x\log x}}$$

is of order 3. As the results stated in the theorem are true of algebraical functions, it is sufficient to prove that, if true of L-functions of order $n-1$, they are true of L-functions of order n.

Let us observe first that if f and ϕ are L-functions, so is f/ϕ. Hence the last part of the theorem is a mere corollary of the first part. Again, the derivative of an L-function of order n is an L-function of order n (or less). Hence it is enough to prove that, if the results stated are true of L-functions of order $n-1$, then an L-function of order n is ultimately continuous and of constant sign, *i.e.* that it is continuous and cannot vanish for a series of values of x increasing beyond limit. For, if this is true of any L-function of order n, it is true of the derivative of any such function; and therefore the function itself is ultimately continuous and monotonic.

Now any L-function of order n can be expressed in the form

$$f_n = A\,\{e\phi_{n-1}^{(1)},\; e\phi_{n-1}^{(2)},\; \ldots,\; e\phi_{n-1}^{(r)},\; l\psi_{n-1}^{(1)},\; \ldots\, l\psi_{n-1}^{(s)},\; \chi_{n-1}^{(1)},\; \ldots\chi_{n-1}^{(t)}\}$$
$$= A\,\{z_1,\; z_2,\; \ldots,\; z_q\},$$

say, where $q = r + s + t$, the functions with suffix $n-1$ are L-functions of order $n-1$, and A denotes an algebraical function: and there is therefore an identical relation

$$F \equiv M_0 f_n^{\,p} + M_1 f_n^{\,p-1} + \ldots + M_p = 0,$$

* See my tract *The integration of functions of a single variable* (No. 2 of this series), pp. 5 *et seq.*, where references to Liouville's original memoirs are given.

where the coefficients are polynomials in $z_1, z_2, \ldots, z_q$. These polynomials are comprised in the class of functions

$$M = \Sigma \rho_{n-1} e \sigma_{n-1} \left(l \tau_{n-1}^{(1)} \right)^{\kappa_1} \left(l \tau_{n-1}^{(2)} \right)^{\kappa_2} \ldots \left(l \tau_{n-1}^{(h)} \right)^{\kappa_h},$$

in which the κ's are positive integers, the number of terms in the summation is finite, and the functions with suffix $n-1$ are again L-functions of order $n-1$. So also are

$$\frac{dM_0}{dx}, \ \frac{dM_1}{dx}, \ \ldots, \ \frac{dM_p}{dx},$$

and the discriminant of F qua function of f_n.

Let us suppose our conclusions established in so far as relates to functions of the type M. Then it follows by a well known theorem* that f_n is continuous, and, since $f_n = 0$ involves $M_p = 0$, that f_n also is ultimately of constant sign.

Hence it is enough to establish our conclusions for functions of the type M. Let us call

$$\kappa_1 + \kappa_2 + \ldots + \kappa_h$$

the *degree* of a term of M, and let us suppose that the greatest degree of a term of M is λ, and that there are μ terms of degree λ, and that the term printed in the expression of M above is one of them.

In the first place it is obvious, from the form of M and the fact that ey and ly are ultimately continuous when y is ultimately continuous and monotonic, that M is ultimately continuous. Again, if M vanishes for values of x surpassing all limit, the same is true of

$$M/(\rho_{n-1} \, e \sigma_{n-1}),$$

and therefore, by Rolle's theorem†, of the derivative of the latter function. But the reader will easily verify that when we differentiate, and arrange the terms of the derivative in the same manner as those of M, we obtain a function of the same form as M but containing at most $\mu - 1$ terms of order λ. And by repeating this process we clearly arrive ultimately at a function of the form

$$N = \Sigma \, \rho_{n-1} \, e \sigma_{n-1},$$

---

* If $F(x, y)$ is a function of x and y which vanishes for $x = a$, $y = b$, and has derivatives $\dfrac{\partial F}{\partial x}$, $\dfrac{\partial F}{\partial y}$ continuous about (a, b), and if $\dfrac{\partial F}{\partial y}$ does not vanish for $x = a$, $y = b$, then there is a unique continuous function y which is equal to b when $x = a$, and satisfies the equation $F(x, y) = 0$ identically. See, *e.g.*, W. H. Young, *Proc. Lond. Math. Soc.*, vol. 7, pp. 397 *et seq.*

† If a function possesses a derivative for all values of its argument, the derivative must have at least one root between any two roots of the function itself.

in which there are no factors of the form $l\tau_{n-1}$, and which must vanish for a sequence of values of x surpassing all limit. Hence it is sufficient for our purpose to prove that this is impossible.

Let the number of terms in N be ϖ. Then

$$\frac{d}{dx}\{N/(\rho_{n-1}\,e\sigma_{n-1})\}$$

must (for reasons similar to those advanced above) vanish for values of x surpassing all limit. But when we differentiate, and arrange the terms of the derivative in the same manner as those of N, we are left with a function of the same form as N, but containing only $\varpi - 1$ terms. And it is clear that a repetition of this process leads to the conclusion that a function of the type

$$\rho_{n-1}\,e\sigma_{n-1}$$

vanishes for values of x surpassing all limit, which is *ex hypothesi* untrue. Hence the theorem is established.

3. The proof just given, it may be observed, does not in any way depend upon the fact that the symbols of algebraical functionality, admitted into the definition of L-functions, are of an *explicit* character. We might admit such functions as

$$e_2\,\surd(ly),$$

where $y^5 + y - x = 0$. But the case contemplated in the definition seems to be the only one of any interest.

Another interesting theorem is: *if f is any L-function, we can find an integer k such that*

$$f \prec e_k x\,;$$

and, if $f \succ 1$, we can find k so that

$$f \succ l_k x:$$

that is to say, an L-function cannot increase more rapidly than any exponential, or more slowly than any logarithm.

More precisely, an L-function of order n cannot satisfy $f \succ e_n(x^{\Delta})$ or $1 \prec f \prec (l_n x)^{\delta}$. The first part of this result is easily established; the second appears to require a more elaborate proof.

4. Let f and ϕ be any two L-functions which tend to infinity with x, and let a be any positive number. Then one of the three relations

$$f \succ \phi^a,\quad f \asymp \phi^a,\quad f \prec \phi^a$$

must hold between f and ϕ; and the second can hold for at most one value of a. If the first holds for any a it holds for any smaller a ; and if the last holds for any a it holds for any greater a.

Then there are three possibilities. Either the first relation holds for every a ; then

$$f \succ \phi^{\Delta}.$$

Or the third holds for every u ; then

$$f \prec \phi^{\delta}.$$

Or the first holds for some values of a and the third for others ; and then there is a value a of a which divides the two classes of values of a, and we may write

$$f = \phi^{a} f_{1},$$

where $\phi^{-\delta} \prec f_{1} \prec \phi^{\delta}$. We shall find this result very useful in the sequel.

IV.

SPECIAL PROBLEMS CONNECTED WITH LOGARITHMICO-EXPONENTIAL SCALES.

1. The functions $e_{r}(l_{s}x)^{\mu}$. We have agreed to express the fact that, however large be a and however small be β, x^{a} has an increase less than that of $e^{x^{\beta}}$, by

$$x^{\Delta} \prec e^{x^{\delta}} \quad\quad\quad\quad\quad\quad (1)^{*}.$$

Let us endeavour to find a function f such that

$$x^{\Delta} \prec f \prec e^{x^{\delta}} \quad\quad\quad\quad\quad (2).$$

If $\phi_{1} \succ \phi_{2}$, $e^{\phi_{1}} \succ e^{\phi_{2}}$ (II. § 8). Thus (2) will certainly be satisfied if

$$\log x \prec \log f \prec x^{\delta}.$$

Hence a solution of our problem is given by

$$f = e^{(\log x)^{1+\delta}}.$$

* Such a relation as

$$x^{\Delta_{1}} (lx)^{\Delta_{2}} \prec e^{\delta_{1} x^{\delta_{2}} (lx)^{-\Delta_{3}}}$$

might at first sight appear to afford more information than (1) : but

$$x^{\Delta_{1}} (lx)^{\Delta_{2}} \prec x^{\Delta_{1}'}, \quad \delta_{1} x^{\delta_{2}} (lx)^{-\Delta_{3}} \succ x^{\delta_{2}'}.$$

where Δ_{1}', δ_{2}' are any positive numbers greater than Δ_{1} and less than δ_{2} respectively. Hence our relation really expresses no more than (1).

Similarly we can prove that

$$f = e^{(\log x)^{1-\delta}}$$

satisfies

$$(\log x)^{\Delta} \prec f \prec x^{\delta}.$$

It will be convenient to write

$$e_0 x \equiv l_0 x \equiv x,$$

and then we have the relations

$$e_0(l_1 x)^{\gamma} \prec e_1(l_1 x)^{1-\delta} \prec e_0(l_0 x)^{\gamma} \prec e_1(l_1 x)^{1+\delta} \prec e_1(l_0 x)^{\gamma} \ldots \ldots (3),$$

where γ denotes *any* positive number*.

Let us now consider the functions

$$f = e_r(l_s x)^{\mu}, \quad f' = e_{r'}(l_{s'} x)^{\mu},$$

where μ, μ' are positive and not equal to 1. If $r = r'$, $f \succ f'$ or $f \prec f'$ according as $s < s'$ or $s > s'$. If $s = s'$, the same relations hold according as $r > r'$ or $r < r'$. If $r = r'$ and $s = s'$, then $f \succ f'$ or $f \prec f'$ according as $\mu > \mu'$ or $\mu < \mu'$. Leaving these cases aside, suppose $s > s'$, $s - s' = \sigma > 0$. Putting $l_{s'} x = y$, we obtain

$$f = e_r(l_{\sigma} y)^{\mu}, \quad f' = e_{r'} y^{\mu'}.$$

If $r < r'$ it is clear that $f \prec \phi$. If $r > r'$, let $r - r' = \rho$; then

$$l_r f = (l_{\sigma} y)^{\mu}, \quad l_r f' = l_{\rho} y^{\mu'} \asymp l_{\rho} y:$$

if $\rho > 1$ the symbol $\asymp$ may be replaced by $\sim$. If $\sigma > \rho$, $l_r f \prec l_r f'$ and so $f \prec f'$. If $\sigma < \rho$, $f \succ f'$. If $\sigma = \rho$, $f \succ f'$ or $f \prec f'$ according as $\mu > 1$ or $\mu < 1$. Thus

$$f \succ f' \ (r - s > r' - s'), \quad f \prec f' \ (r - s < r' - s'),$$

while if $r - s = r' - s'$, $f \succ f'$ or $f \prec f'$ according as $\mu > 1$ or $\mu < 1$, μ being the exponent of the logarithm of higher order which occurs in f or f'.

From this it follows that

$$\ldots e_1(l_2 x)^{1-\delta} \prec e_0(l_1 x)^{\gamma} \equiv (lx)^{\gamma} \prec e_1(l_2 x)^{1+\delta} \prec e_2(l_3 x)^{1+\delta} \prec \ldots$$
$$\ldots \prec e_2(l_2 x)^{1-\delta} \prec e_1(l_1 x)^{1-\delta} \prec e_0(l_0 x)^{\gamma} \equiv x^{\gamma} \prec e_1(l_1 x)^{1+\delta} \prec \ldots$$
$$\ldots \prec e_3(l_2 x)^{1-\delta} \prec e_2(l_1 x)^{1-\delta} \prec e_1(l_0 x)^{\gamma} \equiv ex^{\gamma} \prec e_2(l_1 x)^{1+\delta} \prec \ldots$$

These relations enable us to interpolate to any extent among what we may call the fundamental logarithmico-exponential orders of infinity, viz. $(l_k x)^{\gamma}$, x^{γ}, $e_k x^{\gamma}$. Thus

$$e^{(lx)^{1+\delta}}, \quad e^{e^{(llx)^{1+\delta}}}, \quad \ldots,$$

and

$$e^{e^{(lx)^{1-\delta}}}, \quad e^{e^{e^{(llx)^{1-\delta}}}}, \quad \ldots,$$

are two scales, the first rising from above x^{γ}, the second falling from below ex^{γ}, and never overlapping.

These scales, and the analogous scales which can be interpolated between other pairs of the fundamental logarithmico-exponential orders, possess

---

* Here δ, as usual, denotes ' any positive number however small.' Of course, in using the index $1 - \delta$, it is tacitly implied that $\delta < 1$.

another interesting property. The two scales written above *cover up* (to put it roughly) *the whole interval between* x^γ *and* ex^γ, *so far as L-functions* (III. § 2) *are concerned*: that is to say, it is impossible that an L-function f should satisfy

$$f \succ e_r (l_r x)^{1+\delta}, \qquad \text{(every } r),$$

$$f \prec e_{r+1} (l_r x)^{1-\delta}, \qquad \text{(every } r);$$

and the corresponding pairs of scales lying between $(l_{k+1}x)^\gamma$ and $(l_k x)^\gamma$, or between $e_k x^\gamma$ and $e_{k+1} x^\gamma$, possess a similar property. This property is analogous to that possessed (III. § 3) by the scales $(l_r x)$, $(e_r x)$; viz. that no L-function f can satisfy $f \succ e_r x$, or $1 \prec f \prec l_r x$, for all values of r. A little consideration is all that is needed to render this theorem plausible: to attempt to carry out the details of a formal proof would occupy more space than we can afford.

2. (i) *Compare the rates of increase of*

$$f = (lx)^{(lx)^\mu}, \qquad \phi = x^{(lx)^{-\nu}}.$$

These functions are the same as $e\{(lx)^\mu llx\}$, $e\{(lx)^{1-\nu}\}$. If $\mu + \nu \geqq 1$, $f \succ \phi$; if $\mu + \nu < 1$, $f \prec \phi$.

(ii) *Compare the rates of increase of*

$$f = x^a (lx)^b, \qquad \phi = e^{A(lx)^a (llx)^\beta}, \qquad (a, A, a > 0).$$

Here $f = e(alx + bllx)$. If $a < 1$, then $f \succ \phi$; if $a > 1$, then $f \prec \phi$. If $a = 1$, $\beta < 0$, then $f \succ \phi$; if $a = 1$, $\beta > 0$, then $f \prec \phi$. If $a = 1$, $\beta = 0$, $a > A$, then $f \succ \phi$; if $a = 1$, $\beta = 0$, $a < A$, then $f \prec \phi$. If $a = 1$, $\beta = 0$, $a = A$, then $f \succ \phi$ if $b > 0$ and $f \prec \phi$ if $b < 0$. Finally if $a = 1$, $\beta = 0$, $a = A$, $b = 0$ the two functions are identical.

(iii) *Compare the increase of* $f = x^{\phi/(1+\phi)}$, *where* ϕ *is a function of* x *such that* $\phi \succ 1$, *with that of* x^γ.

It is clear that $f \preccurlyeq x$, but $f \succ x^\gamma$ for any value of γ less than unity. For, if x is large enough, $\phi > n$, where n is any positive integer, and so

$$f > x^{n/(1+n)}.$$

Again $f = xe^{-lx/(1+\phi)}$, and so, if $\phi \prec lx$, $f \prec x$: but if $\phi \asymp lx$, $f \asymp x$; while if $\phi \succ lx$, $f \sim x$.

3. Successive approximations to a logarithmico-exponential function. Consider such a function as

$$f = \sqrt{(x)}\,(lx)^2\, e^{\sqrt{(lx)}\,(l_2 x)^2}\, e^{\sqrt{(l_2 x)}\,(l_3 x)^2}.$$

If we omit one or more of the parts of the expression of f we obtain another function whose increase differs more or less widely from that of f. The question arises as to which parts are of the greatest and which of the least importance; *i.e.* as to which are the parts whose omission affects the increase of f most or least fundamentally.

Taking logarithms we find

$$lf = \tfrac{1}{2} lx + \sqrt{(lx)}\,(l_2 x)^2\, e^{\sqrt{(l_2 x)}\,(l_3 x)^2} + 2l_2 x \quad \dots\dots\dots\dots(a),$$

the three terms being arranged in order of importance. Again

$$l_2 f = l_2 x - l2 + \epsilon, \quad l_3 f = l_3 x + \epsilon,$$

where (I. § 5) in each of the last equations ϵ denotes a function (not the same function) which tends to zero as $x \to \infty$. If we neglect this term in each of them in turn we deduce the approximations

$$(1) \ f = x, \quad (2) \ f = \sqrt{x}.$$

By neglecting the last term in the equation (a) we obtain the much closer approximation

$$(6) \quad f = \sqrt{(x)} \, e^{\sqrt{(lx)(l_2 x)^2}} e^{\sqrt{(l_2 x)(l_3 x)^2}}.$$

In order to obtain a more complete series of approximations to f we must replace the equation (a) by a series of approximate equations. Now if

$$\phi = \sqrt{(lx)(l_2 x)^2} \, e^{\sqrt{(l_2 x)(l_3 x)^2}}$$

we have
$$l\phi = \tfrac{1}{2} l_2 x + \sqrt{(l_2 x)(l_3 x)^2} + 2l_3 x,$$

$$l_2 \phi = l_3 x - l2 + \epsilon, \quad l_3 \phi = l_4 x + \epsilon.$$

Hence we obtain (0) $\phi = lx$, (3) $\phi = \sqrt{(lx)}$, and (5) $\phi = \sqrt{(lx)} \, e^{\sqrt{(l_2 x)(l_3 x)^2}}$ as approximations to the increase of ϕ: of these, however, the first is valueless, inasmuch as it would make ϕ preponderate over the first term on the right hand side of (a).

A similar argument, applied to the function $e^{\sqrt{(l_2 x)(l_3 x)^2}}$, leads us to interpolate (4) $\phi = \sqrt{(lx)} \, e^{\sqrt{(l_2 x)}}$ between (3) and (5). We can now, by adopting a series of approximate forms of the equation (a), deduce a complete system of closer and closer approximations to the increase of f, viz.

$$(1) \ x, \quad (2) \ \sqrt{x}, \quad (3) \ \sqrt{(x)} \, e^{\sqrt{(lx)}}, \quad (4) \ \sqrt{(x)} \, e^{\sqrt{(lx)}} e^{\sqrt{(l_2 x)}},$$

$$(5) \ \sqrt{(x)} \, e^{\sqrt{(lx)}} e^{\sqrt{(l_2 x)(l_3 x)^2}}, \quad (6) \ \sqrt{(x)} \, e^{\sqrt{(lx)}(l_2 x)^2} e^{\sqrt{(l_2 x)(l_3 x)^2}}.$$

This order corresponds exactly to the order of importance of the various parts of the expression of f.

4. Legitimate and illegitimate forms of approximation to a logarithmico-exponential function.

In applications of this theory, such as occur, for instance, in the theory of integral functions, we are continually meeting such equations as

$$f = (1 + \epsilon) e^{x^a}, \quad f = e^{(1 + \epsilon) x^a}, \quad f = e^{x^{a + \epsilon}}, \quad (a > 0) \ldots \ldots (1).$$

It is important to have clear ideas as to the degree of accuracy of such representations of f. The simplest method is to take logarithms repeatedly, as in § 3 above.

In the first example the term ϵ does not affect the increase of f: we have $f \sim e x^a$. This is not true in the second; but $lf \sim x^a$, so that the term ϵ does not affect the increase of lf; while in the third this is not true, though $llf \sim a$. Of the three formulae the first gives the most, and the last the least, information as to the increase of f (see also VII. § 3).

Such a formula as

$$f = x e^{(1 + \epsilon)} x^{\alpha} \quad \dots \dots \dots \dots \dots \dots \dots \dots \dots (2)$$

would not be a legitimate form of approximation at all. For the factor $e(\epsilon x^{\alpha})$ which is not completely specified may well be far more important than the explicitly expressed factor x: we might for example have $\epsilon = x^{-\beta}$, where $0 < \beta < a$, in which case $e(\epsilon x^{\alpha})$ is more important than any power of x. Thus (2) does not really convey more information than the second equation (1), and to use it would involve a logical error similar to that involved in saying that the sun's distance is 92,713,600 miles, with a probable error of some 100,000 miles.

5. Attempts to represent orders of infinity by symbols. It is natural to try to devise some simple method of representing orders of infinity by symbols which can be manipulated according to laws resembling as far as possible those of ordinary algebra. Thus Thomae* has proposed to represent the order of infinity of $f = x^{\alpha} (l x)^{\alpha_1} (l_2 x)^{\alpha_2} \dots$ by

$$Of = a + a_1 l_1 + a_2 l_2 + \dots \dagger,$$

where the symbols l_1, l_2, ... are to be regarded as new units. It is clear that these units cannot, in relation to one another, obey the Axiom of Archimedes ‡: however great n, $n l_2$ cannot be as great as l_1, nor $n l_1$ as great as 1.

The consideration of a few simple cases is enough to show that any such notation, if it is to be of any use, must obey the following laws :

(i)   if $f \succcurlyeq \phi$, $O(f + \phi) = Of$;

(ii) $O(f\phi) = Of + O\phi$;

(iii) $O\{f(\phi)\} = Of \times O\phi$.

And Pincherle§ has pointed out that these laws are in any case inconsistent with the maintenance of the laws of algebra in their entirety. Thus if

$$Ox = 1, \quad Olx = \lambda,$$

we have, by (iii), $Ollx = \lambda^2$, and by (iii) and (ii),

$$Ol(xlx) = \lambda (1 + \lambda) = \lambda + \lambda^2 ;$$

and on the other hand, by (i),

$$Ol(xlx) = O(lx + llx) = \lambda.$$

Pincherle has suggested another system of notation ; but the best yet formulated is Borel's‖.   Borel preserves the three laws (i), (ii), (iii), the

---

* *Elementare Theorie der analytischen Funktionen*, S. 112.

† The reader will not confuse this use of the symbol O (which does not extend beyond this paragraph) with that explained in I. § 5.

‡ 'If $x > y > 0$, we can find an integer n such that $ny > x$.'

§ *l.c.* (see p. 13 above).

‖ *Leçons sur les séries à termes positifs*, pp. 35 et seq. ; for further information see his recently published *Leçons sur la théorie de la croissance*, pp. 14 et seq.

commutative law of addition, and the associative law of multiplication. But multiplication is no longer commutative, and only distributive on one side*.

He would denote the orders of

$$e^x x^n, \quad x^n (lx)^p, \quad e^{2x}, \quad e^{x^2}, \quad e^{e^x}, \quad e^{\sqrt{lx}}, \quad \tfrac{1}{2}x,$$

by

$$\omega + n, \quad n + \frac{p}{\omega}, \quad 2.\omega, \quad \omega.2, \quad \omega^2, \quad \omega . \frac{1}{2}.\frac{1}{\omega}, \quad \frac{1}{\omega}.\frac{1}{2}.\omega.$$

But little application, however, has yet been found for any such system of notation; and the whole matter appears to be rather of the nature of a mathematical curiosity.

V.

FUNCTIONS WHICH DO NOT CONFORM TO ANY LOGARITHMICO-EXPONENTIAL SCALE.

1. WE saw in I. (§ 2) that, given two increasing functions ϕ and ψ ($\phi \succ \psi$), we can always construct an increasing function f which is, for an infinity of values of x increasing beyond all limit, of the order of ϕ, and for another infinity of values of x of the order of ψ. The actual construction of such functions by means of explicit formulae we left till later. We shall now consider the matter more in detail, with special reference to the case in which ϕ and ψ are L-functions.

We shall say that f is an *irregularly increasing* function (*fonction à croissance irrégulière*) if we can find two L-functions ϕ and ψ ($\phi \succ \psi$) such that

$$f \geqslant \phi \quad (x = x_1, x_2, \ldots), \quad f \leqslant \psi \quad (x = x_1', x_2', \ldots),$$

$x_1, x_2, \ldots$ and $x_1', x_2', \ldots$ being any two indefinitely increasing sequences of values of x. We shall also say that 'the increase of f is irregular' and that 'the logarithmico-exponential scales are *inapplicable* to f.'

The phrase '*fonction à croissance irrégulière*' has been defined by various writers in various senses. Borel† originally defined f to be *à croissance régulière* if

$$e^{x^{a-\delta}} < f < e^{x^{a+\delta}}, \qquad (x > x_0),$$

or in other words if $llf \sim alx$ or $llf \gtrless lx$.

This definition was of course designed to meet the particular needs of the

---

* $(a + b) c = ac + bc$, but in general $a (b + c) \neq ab + ac$.

† *Leçons sur les fonctions entières*, p. 107.

theory of integral functions: and has been made more precise by Boutroux and Lindelöf*, who use inequalities of the form

$$e^{x^\alpha\,(lx)^{\alpha_1}\,\dots\,(l_k x)^{\alpha_k-\delta}} < f < e^{x^\alpha\,(lx)^{\alpha_1}\,\dots\,(l_k x)^{\alpha_k+\delta}}.$$

All functions which are not *à croissance régulière* for these writers are included in our class of irregularly increasing functions.

2. The logarithmico-exponential scales may fail to give a complete account of the increase of a function in two different ways. The function may be of irregular increase, as explained above, and the scales *inapplicable*: on the other hand they may be, not inapplicable, but *insufficient* (*en défaut*). That is to say, although the increase of the function does not oscillate from that of one L-function to that of another, there may be no L-function capable of measuring it. That such functions exist follows at once from the general theorems of II. Thus we can define a function which tends to infinity more rapidly than any $e_r x$, or more slowly than any $l_r x$: and the increase of such a function is more rapid or slower than that of any L-function (III. § 2). Or again, we can (II. § 6) define a function whose increase is greater than that of $e_r(l_r x)^{1+\delta}$ (any r) and less than that of $e_{r+1}(l_r x)^{1-\delta}$ (any r); and the increase of such a function (IV. § 1) cannot be equal to that of any L-function.

We shall now discuss some actual examples of functions for which the logarithmico-exponential scales are inapplicable or insufficient.

3. Irregularly increasing functions. Functions whose increase is irregular may be constructed in a variety of ways.

(i) Pringsheim† has used, in connection with the theory of the convergence of series, functions of an integral variable n whose increase is irregular. A simple example of such a function is

$$f(n) = 10^{[(\log_{10} n)^{1/\tau}]^\tau}, \qquad (\tau > 1),$$

where $[x]$ denotes the integral part of x. It is easily proved, for instance, when $\tau = 2$, that the increase of $f(n)$ varies between that of n and that of $n \cdot 10^{1-2\sqrt{(\log_{10} n)}}$. We shall not do more than mention functions of this type. They are defined, most naturally, as functions of an integral variable n: if we extend the definition to the continuous variable, the resulting function is discontinuous. The definition can of course be modified so as to give a

---

* Boutroux, *Acta Mathematica*, t. 28, p. 97; Lindelöf, *Acta Societatis Fennicae*, t. 31, p. 1. See also Blumenthal, *Principes de la théorie des fonctions entières d'ordre infini*.

† See *Math. Annalen*, Bd. 35, S. 347 *et seq.* and *Münchener Sitzungsberichte*, Bd. 26, S. 605 *et seq.*

continuous function of x with substantially the same properties; but it is not easy to effect this by a simple, natural, and explicit formula.

(ii)   A more natural type of function is given by

$$f = \phi \cos^2 \theta + \psi \sin^2 \theta,$$

where ϕ, ψ, θ are increasing L-functions. We have to consider what conditions ϕ, ψ, θ must satisfy in order that f may increase steadily with x. That its increase oscillates between that of ϕ and that of ψ is obvious.
Differentiating,

$$f' = \phi' \cos^2 \theta + \psi' \sin^2 \theta + 2(\psi - \phi)\theta' \cos\theta \sin\theta.$$

Suppose $\phi \succ \psi$: and let us assume that (as will be proved in the next chapter) relations between L-functions involving the symbols $\succ$, etc. may be differentiated and integrated. The condition that f' should always be positive is $\phi'\psi' \succ (\phi - \psi)^2 \theta'^2$ or $\phi'\psi' \succ \phi^2 \theta'^2$. A fortiori, since $\phi' \succ \psi'$, we must have $\phi' \succ \phi\theta'$, or $\log\phi \succ \theta$. Thus f is certainly monotonic if

$$\phi \succ \psi, \quad \log\phi \succ \theta, \quad \psi' \succ \phi^2 \theta'^2/\phi'.$$

If, e.g., $\theta = x$, we require $\log\phi \succ x$, which is satisfied, for example, if $\phi = x^a e^{x^\rho}$ $(\rho > 1)$. It is convenient to write $a + \rho - 1$ for u. Then, since $\phi' \sim \rho x^{a+\rho-1} e^{x^\rho}$, we must have $\psi' \succ x^a e^{x^\rho}$; and so

$$\psi \succ \int^x t^a e^{t^\rho} dt = \frac{1}{\rho} \int^x t^{a-\rho+1} \frac{d}{dt}(e^{t^\rho}) dt \sim \frac{1}{\rho} x^{a-\rho+1} e^{x^\rho},$$

as is easily seen on integrating by parts. Thus we may take $\psi = x^\beta e^{x^\rho}$, where $a - 2\rho + 2 < \beta < a$. Changing our notation a little we see that

$$f = (x^{\gamma+\delta} \cos^2 x + x^{\gamma-\delta} \sin^2 x) e^{x^\rho}$$

is monotonic if $0 < \delta < \rho - 1$; and the increase of f obviously oscillates between that of $x^{\gamma+\delta} e^{x^\rho}$ and that of $x^{\gamma-\delta} e^{x^\rho}$. Similarly it may be shown that

$$f = (e^{\mu x} \cos^2 x + e^{\nu x} \sin^2 x) e e^x$$

is monotonic if $\nu < \mu < \nu + 2$*; and again the increase of f is irregular.

4.   Irregularly increasing functions (*continued*). We shall now consider two more general and more important methods for the construction of irregularly increasing functions.

(iii)   Borel† has shown how, by means of power series, we may define functions which increase steadily with x, while their increase oscillates to an arbitrary extent.

---

* Cf. *Messenger of Mathematics*, vol. 31, p. 1.

† See Borel, *Leçons sur les fonctions entières*, pp. 120 et seq.; *Leçons sur les séries à termes positifs*, pp. 32 et seq. Borel considers the cases only in which $\psi = e^x$, $\phi = e^{x^2}$ or e^{e^x}; but his method is obviously of general application. The proof here given is however more general and much simpler.

Let
$$\phi(x) = \Sigma a_n x^n, \quad \psi(x) = \Sigma b_n x^n$$

be two integral functions of x with positive coefficients; suppose also $\phi \succ \psi$. The increase of ϕ and ψ may be as large as we like (II. § 4); but in each case it must be greater than that of any power of x.

Then we can define a function

$$f(x) = \Sigma c_n x^n,$$

where every c_n is equal either to a_n or to b_n, in such a way that, for an infinity of values x_ν whose limit is infinity, we have $f \sim \phi$, and for a similar infinity of values x_ν' we have $f \sim \psi *$.

Let (η_ν) be a sequence of decreasing positive numbers whose limit is zero. Take a positive number x_0 such that $\phi(x_0) > 1$, $\psi(x_0) > 1$, and a number x_1 greater than x_0. When x_1 is fixed we can choose n_1 so that

$$\sum_{n_1}^{\infty} a_n x_1^n < \tfrac{1}{3}\eta_1, \quad \sum_{n_1}^{\infty} b_n x_1^n < \tfrac{1}{3}\eta_1,$$

and so, if c_n is either of a_n, b_n (however the selection may be made for different values of n),

$$\sum_{n_1}^{\infty} c_n x_1^n < \sum_{n_1}^{\infty} (a_n + b_n) x_1^n < \tfrac{2}{3}\eta_1.$$

For $0 \leqslant n < n_1$ we take $c_n = a_n$. Then

$$| f(x_1) - \phi(x_1) | < \sum_{n_1}^{\infty} (a_n + c_n) x_1^n < \eta_1,$$

and so, since $\phi(x_1) > 1$,

$$\left| \frac{f(x_1)}{\phi(x_1)} - 1 \right| < \eta_1 \quad\quad\quad\quad\quad\quad (1).$$

Now let x_2 be a number greater than x_1; we can suppose x_2 chosen so that

$$\left(\sum_{0}^{n_1-1} a_n x_2^n \right) \Big/ \psi(x_2) < \tfrac{1}{5}\eta_2, \quad \left(\sum_{0}^{n_1-1} b_n x_2^n \right) \Big/ \psi(x_2) < \tfrac{1}{5}\eta_2.$$

When x_2 is fixed we can choose n_2 ($n_2 > n_1$) so that

$$\sum_{n_2}^{\infty} a_n x_2^n < \tfrac{1}{5}\eta_2, \quad \sum_{n_2}^{\infty} b_n x_2^n < \tfrac{1}{5}\eta_2.$$

For $n_1 \leqq n < n_2$ we take $c_n = b_n$. And, however c_n be chosen for $n \geqq n_2$, we have

$$\sum_{n_2}^{\infty} c_n x_2^n < \sum_{n_2}^{\infty} (a_n + b_n) x_2^n < \tfrac{2}{5}\eta_2.$$

* By '$f \sim \phi$ for an infinity of values x_ν' we mean of course that $f/\phi \to 1$ as $x \to \infty$ through this particular sequence of values.

Also

$$|f(x_2) - \psi(x_2)| < \sum_0^{n_1-1} a_n x_2{}^n + \sum_0^{n_1-1} b_n x_2{}^n + \sum_{n_2}^{\infty} c_n x_2{}^n + \sum_{n_2}^{\infty} b_n x_2{}^n$$

$$< \tfrac{2}{5}\eta_2 \psi(x_2) + \tfrac{3}{5}\eta_2 < \eta_2 \psi(x_2),$$

and so

$$\left| \frac{f(x_2)}{\psi(x_2)} - 1 \right| < \eta_2 \quad\dots\dots\dots\dots\dots(2).$$

It is plain that, by a repetition of this process, we can find a sequence $x_1, x_2, x_3, \dots$ whose limit is infinity, so that

$$\left| \frac{f(x_3)}{\phi(x_3)} - 1 \right| < \eta_3 \dots\dots(3), \qquad \left| \frac{f(x_4)}{\psi(x_4)} - 1 \right| < \eta_4 \dots\dots(4), \qquad \dots ;$$

and our conclusion is thus established. Incidentally we may remark that not only f itself, but all its derivatives also, are increasing and continuous.

It is clear that, if we were given any number of integral functions $\phi_1, \phi_2, \dots, \phi_k$, with positive coefficients, we could define f so that $f/\phi_s \to 1$, as $x \to \infty$ through a suitably chosen sequence of values, for each of the functions ϕ_s.

(iv) **Power series with gaps.** There is another method of constructing irregularly increasing functions by means of power series which, though less general theoretically than that explained above, is in some ways more interesting, inasmuch as the functions to which it leads us are of a far simpler and more natural type. We shall confine ourselves here to explaining in general terms the general principle of the method and indicating a few simple examples*.

Let

$$\phi(x) = \Sigma a_n x^n \quad\dots\dots\dots\dots\dots(1)$$

be an integral function with positive coefficients: suppose, to fix our ideas, that the coefficients decrease steadily as n increases. Suppose also that, for a particular value of x,

$$\varpi(x) = a_\nu x^\nu$$

is the greatest term of the series. In general one term will be the greatest, but for certain particular values of x, say $\xi_1, \xi_2, \dots$, two consecutive terms will be equal†.

---

* For fuller details see Hardy, *Proc. Lond. Math. Soc.*, vol. 2, pp. 332 *et seq.*; *Messenger of Mathematics*, vol. 39, p. 28: Borel, *Rendiconti del Circolo Matematico di Palermo*, t. 23, p. 320; *Leçons sur la théorie de la croissance*, pp. 111 *et seq.*: Blumenthal, *Principes de la théorie des fonctions entières d'ordre infini*, pp. 5 *et seq.*

† We leave aside the possibility, which obviously applies only to particular cases, of more than two terms being equal.

As x increases, the index ν of $\varpi(x)$ increases, and tends to ∞ with n: it thus defines a function $\nu(x)$ such that

$$\nu(x) = i, \quad (\xi_i < x < \xi_{i+1}).$$

At the point of discontinuity ξ_i, where $\nu(x)$ jumps from $i-1$ to i, we may assign to it the value i. When ν is thus defined for all values of x, $\varpi(x)$ defines a function of x which tends continuously and steadily to ∞ with x.

The increase of ϕ is obviously at least as great as that of ϖ; it may be expected to be greater: but it is, in ordinary cases, not so very much greater—the increase of ϖ gives a very fair approximation to that of ϕ. Thus, if $\phi(x) = e^x$, $a_n = 1/n!$, and $\xi_i = i$. And for $i < x < i+1$ we have

$$e^i < \phi < e^{i+1}, \quad (1-\epsilon_i)\,\frac{e^i}{\sqrt{(2\pi i)}} < \varpi < (1+\epsilon_i)\,\frac{e^{i+1}}{\sqrt{(2\pi i)}}*.$$

Thus $\phi \succ \varpi$, but $\log \phi \sim \log \varpi$: the difference between the increases of ϕ and ϖ is small compared with the increases themselves.

Now let

$$f(x) = \Sigma\, a_{\chi(n)}\, x^{\chi(n)} \quad\dots\dots\dots\dots\dots(2),$$

where $\chi(n) \succ n$: and let $p(x)$ be the function related to f as $\varpi(x)$ is to ϕ. The laws of increase of $\varpi(x)$ and of $p(x)$ may be expected to be very much the same, for $p(x)$ is defined by a selection from *some* of the terms from *all* of which $\varpi(x)$ was selected. The increase of $f(x)$ clearly cannot be greater, and may be expected to be less, than that of $\phi(x)$: but it cannot be less than that of $p(x)$. Hence we may expect relations of the type

$$p \asymp \varpi \prec f \prec \phi\,\dagger.$$

Also it is clear that, the more rapidly we suppose $\chi(n)$ to increase, the lower in the gap between ϖ and ϕ will f sink, and that, if we suppose χ to increase with sufficient rapidity, we may expect to find $\varpi \asymp f$, so that the increase of f is completely dominated by that of one (variable) term.

We then shall have

$$f(x) \asymp a_{N(x)}\, x^{N(x)},$$

where $N(x)$ is a function of x which assumes successively each of a series of integral values N_i, so that

$$N(x) = N_i, \quad (x_i \leqq x < x_{i+1})\,\ddagger.$$

But, as x increases from x_i to x_{i+1}, the order of $a_{N_i} x^{N_i}$, considered as a function of x, may vary considerably, since N_i, though depending on the interval (x_i, x_{i+1}), does not depend on the particular position of x in that interval. And so it is clear that we are in this way likely to be led to functions whose increase is irregular in the sense explained in § 1.

---

* The second pair of inequalities are an immediate consequence of Stirling's theorem, that $i! \sim i^{i+\frac{1}{2}}\, e^{-i}\, \sqrt{(2\pi)}$.

† We *must* have $p \leqslant \varpi$, $p \leqslant f$, $\varpi \leqslant \phi$, $f \leqslant \phi$.

‡ N_i, x_i are, of course, not the same as ν_i, ξ_i above.

Suppose, for example, that $a_n = n^{-n}$, so that

$$\phi(x) = \Sigma \left(\frac{x}{n}\right)^n \sim \sqrt{\left(\frac{2\pi x}{e}\right)}\, e^{x/e}\, *.$$

Here
$$\xi_i = i\left(1 + \frac{1}{i}\right)^{i+1} \sim ei,$$

and it is easily shown that $\varpi(x) \asymp e^{x/e}$.

Now let $\chi(n) = 2^n$, so that

$$f(x) = \Sigma \frac{x^{2^n}}{2^{n2^n}} = \Sigma v_n$$

say. Then $v_{i-1} = v_i$ if $x = 2^{i+1}$, so that $x_i = 2^{i+1}$ and $N_i = 2^i$ for

$$2^{i+1} \leqq x < 2^{i+2}.$$

For this range of values of x, v_i is the greatest term; when $x = 2^{i+2}$, $v_i = v_{i+1}$.
Further, it is not difficult to show that $f(x) \asymp p(x) = v_i$, the behaviour of
$f(x)$ being dominated by that of its greatest term†.

If we put $x = 2^{i+1+\theta}$, where $0 < \theta < 1$, we find

$$f(x) \asymp v_i = 2^{(1+\theta)2^i} = 2^{ax},$$

where $a = (1+\theta)2^{-1-\theta}$. This is a maximum when $1 + \theta = 1/(\log 2)$, when it
is equal to $1/(e \log 2) = \cdot 53\ldots$ Hence the increase of $f(x)$ oscillates (roughly)
between that of $2^{\cdot 53\ldots x}$ and $2^{\frac{1}{2}x+1}$‡.

Similar considerations may be applied to the more general series

$$\Sigma \frac{x^{a^n}}{b^{na^n}},$$

where a is an integer greater than unity. This series is derived from $\Sigma(x/n^a)^n$,
where $a = (\log b)/(\log a)$, by taking $\chi(n) = a^n$. Another example of an irregularly increasing function defined in a similar manner is

$$f(x) = \Sigma \frac{x^{n^3}}{(n^3)!},$$

the increase of which oscillates between the increases of $e^x/\sqrt{x}$ and

$$x^{-\frac{1}{2}} e^{x - \frac{9}{8}x^{1/3}}\S.$$

These examples are of course typical of a large class of functions.

Before we leave this subject let us call attention to a point of considerable

---

* See II. § 3, and the references given in the footnote to p. 10. We might
have taken $\phi(x) = e^x$, but our choice of $\phi(x)$ leads to the simplest examples.

† We may say roughly that *in general* $f \sim p$—that is to say, $f/p \to 1$ as $x \to \infty$
through any sequence of values not falling inside any of certain intervals sur-
rounding the values ξ_i. At a point ξ_i, f/p is nearly equal to 2.

‡ The latter function is multiplied by 2, as there are two equal terms when
$\theta = 0$ or 1.

§ *Messenger of Mathematics*, vol. 39, p. 28.

interest suggested by the foregoing examples. In forming the logarithmico-exponential scales we started from the scale x, x^2, ... and then formed the function $\Sigma \dfrac{x^n}{n!}$. If we had started, as we equally well might have done, from the scale x^2, x^4, x^8, ... (cf. II. § 1), we should have been led to choose, as a function transcending this scale, not e^x but some such function as

$$\Sigma \frac{x^{2^n}}{(2^n)!}.$$

This is one of the irregularly increasing functions of the type just considered. Had we proceeded thus, and completed the construction of our fundamental scales on similar lines, our fundamental functions would for the most part have been among those which do not conform to the logarithmico-exponential scale, and it would have been the functions of that scale that would have appeared as irregularly increasing functions.

5. Functions which transcend the logarithmico-exponential scales. We turn our attention now to functions for which the logarithmico-exponential scales are not inapplicable but *insufficient* (§ 2). Of the existence of such functions we are already assured. Thus a function which assumes the values $e_1(1)$, $e_2(2)$, ..., $e_\nu(\nu)$, ... for $x = 1, 2, ..., \nu, ...$ certainly has an increase greater than that of any logarithmico-exponential function. No such function, however, has as yet made its appearance naturally in analysis; it will be sufficient, therefore, to mention two examples of such functions which transcend the logarithmico-exponential scales in quite different manners.

(i) The series $\qquad \Sigma \dfrac{e_\nu(x)}{e_\nu(\nu)}$

has obviously, if it converges, an increase greater than that of any $e_\nu(x)$. Suppose $k - 1 \leqq x < k$. Then

$$\frac{e_k(x)}{e_k(k)} < 1, \quad \frac{e_{k+\nu}(x)}{e_{k+\nu}(k+\nu)} < \frac{e_{k+\nu}(k)}{e_{k+\nu}(k+\nu)} < \frac{e_{k+\nu}(k)}{e_{k+\nu}(k+1)}.$$

But, by the Mean Value Theorem,

$$e_{k+\nu}(k+1) = e_{k+\nu}(k) + e_{k+\nu}(y)\, e_{k+\nu-1}(y) \cdots e_2(y)\, e_1(y),$$

where y is some number between k and $k+1$; and so

$$e_{k+\nu}(k+1) > e_{k+\nu}(k)\, e_{k+\nu-1}(k) \cdots e_1(k).$$

It follows that the terms of the series

$$\sum_{\nu=k}^{\infty} \frac{e_\nu(x)}{e_\nu(\nu)}$$

are less than those of the series

$$1 + \sum_{\nu=1}^{\infty} \frac{1}{e_1(k)\, e_2(k) \cdots e_{k+\nu-1}(k)},$$

H.

3

which is plainly convergent, and therefore that the original series is convergent; and it is obviously only one of a large class of series possessing similar properties.

(ii) Let $\phi(x)$ be an increasing function such that $\phi(0) > 0$, $\phi \succ x$. We can define an increasing function f, which satisfies the equation

$$ff(x) = \phi(x) \quad \dots\dots\dots\dots\dots\dots(1),$$

as follows.

Draw the curves $y = x$, $y = \phi(x)$ (Fig. 5). Take Q_0 arbitrarily on OP_0 (see the figure); draw $Q_0 R_1$ parallel to OX and complete the rectangle $Q_0 Q_1$. Join Q_0, Q_1 by any continuous arc everywhere inclined at an acute angle to the axes. On this arc take any point Q; draw QP, QR parallel to the axes, and complete the rectangle QQ'. As Q moves from Q_0 to Q_1, Q' moves from Q_1 to Q_2, say. As we constructed Q' from Q, so we can construct Q'' from Q': proceeding thus we define a continuous curve $Q_0 Q_1 Q_2 Q_3 \dots$ corresponding to a continuous and increasing function $f(x)$. Then $f(x)$ satisfies (1). For if $y = f(x)$

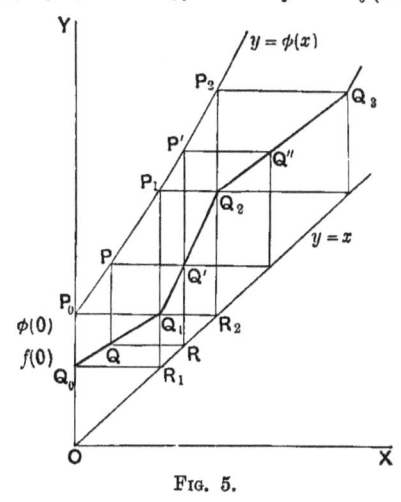

FIG. 5.

is the ordinate of Q, it is clear that $ff(x)$ is the ordinate of Q', which is equal to $\phi(x)$, the ordinate of P.

Let us write

$$f(x) = f_1(x), \quad \phi(x) = f_1 f_1(x) = f_2(x), \quad f\phi(x) = \phi f(x) = f_3(x),$$

and so on, so that Q_n is the point $f_n(0)$, $f_{n+1}(0)$. Also let ψ be the function inverse to ϕ, and write ψ_2 for $\psi\psi$, and so on. Finally, let the equation of $Q_0 Q_1$ be $\theta(x, y) = 0$. Then it is easy to see that the equations of $Q_{2n} Q_{2n+1}$ and of $Q_{2n+1} Q_{2n+2}$ are respectively

$$\theta\{\psi_n(x), \ \psi_n(y)\} = 0, \quad \theta\{\psi_{n+1}(y), \ \psi_n(x)\} = 0.$$

Suppose for example that $\phi(x) = e^x$, $OQ_0 = a < 1$, and that $Q_0 Q_1$ is the straight line $y = a + ax$, where $a = (1-a)/a$. Then the equations of $Q_{2n} Q_{2n+1}$ and of $Q_{2n+1} Q_{2n+2}$ are

$$l_n y = a + a l_n x, \quad l_n x = a + a l_{n+1} y,$$

or

$$y = e_{n-1}\{e^a (l_{n-1} x)^a\}, \quad y = e_n\{e^{-a/a} (l_{n-1} x)^{1/a}\}.$$

For simplicity let us take $a = \frac{1}{2}$, $a = 1$. Then the equations of $Q_{2n} Q_{2n+1}$ and of $Q_{2n+1} Q_{2n+2}$ are respectively

$$y = e_{n-1} \{\sqrt{e} \, (l_{n-1} x)\} = e_{n-2} \{(l_{n-2} x)^{\sqrt{e}}\} \; = \lambda_n(x),$$
$$y = e_n \{(l_{n-1} x)/\sqrt{e}\} \; = e_{n-1} \{(l_{n-2} x)^{1/\sqrt{e}}\} = \mu_n(x),$$

say. Now (IV. § 1)

$$x^\gamma \prec \lambda_3 \prec \dots \prec \lambda_n \prec \dots \prec \mu_n \prec \dots \prec \mu_3 \prec e^{x\gamma}$$

and a function f, such that $\lambda_n \prec f \prec \mu_n$ for all values of n, transcends the logarithmico-exponential scales. But f clearly satisfies these relations, and so its increase is incapable of exact measurement by these scales.

It is easily verified that $\lambda_n \lambda_n x \prec e^x$ and $\mu_n \mu_n x \succ e^x$ for all values of n. Hence it is clear *a priori* that any increasing solution of (1) must satisfy $\lambda_n \prec f \prec \mu_n$.

This kind of 'graphical' method may also be employed to define functions whose increase, like that of the function considered under (i) above, is slower than that of any logarithm or more rapid than that of any exponential. It can be employed, for example, to solve the equation

$$\phi(2^x) = 2\phi(x);$$

and it can be proved that the increase of a function such that $\phi(2^x) \asymp \phi(x)$ is slower than that of any logarithm (VII. § 3).

6. The importance of the logarithmico-exponential scales.

As we have seen in the earlier paragraphs of this section, it is possible, in a variety of ways, to construct functions whose increase cannot be measured by any L-function. It is none the less true that no one yet has succeeded in defining a mode of increase genuinely independent of all logarithmico-exponential modes. Our irregularly increasing functions oscillate, according to a logarithmico-exponential law of oscillation, between two logarithmico-exponential functions; the functions of § 5 were constructed expressly to fill certain gaps in the logarithmico-exponential scales. No function has yet presented itself in analysis the laws of whose increase, in so far as they can be stated at all, cannot be stated, so to say, in logarithmico-exponential terms.

It would be natural to expect that the arithmetical functions which occur in the theory of the distribution of primes might give rise to genuinely new modes of increase. But, so far as analysis has gone, the evidence is the other way.

Thus if we denote by $\varpi(x)$ the number of prime numbers less than x, it is known that

$$\varpi(x) \sim \frac{x}{\log x}.$$

More precisely

$$\varpi(x) = \int_2^x \frac{dt}{\log t} + \rho(x) = Li(x) + \rho(x),$$

where $|\rho(x)| \prec x(\log x)^{-\Delta}$. The precise order of $\rho(x)$ has not yet been determined, but there is reason to anticipate that $\rho(x) \preccurlyeq \sqrt{x}/(\log x)$.

VI.

DIFFERENTIATION AND INTEGRATION.

1. Integration. It is important to know when relations of the types $f(x) \succ \phi(x)$, etc., can be differentiated or integrated. The results are very much what might be expected from analogy with similar results in other branches of analysis, and may therefore be discussed somewhat summarily. For brevity we denote

$$\int_a^x f(t)\, dt, \quad \int_a^x \phi(t)\, dt$$

(where a is a constant) by $F(x)$ and $\Phi(x)$. And we suppose for the moment that f and ϕ are positive for $x \gtrless a$.

It may be well to repeat (cf. I. § 4) that f and ϕ are always supposed to be (at any rate for $x > x_0$) positive, continuous, and monotonic, unless the contrary is stated or clearly implied. Some of our conclusions are valid under more general conditions; but the case thus defined, and the corresponding case in which f or ϕ or both of them are negative, are the only cases of importance.

Lemma. *If $\Phi \succ 1$, and $f > H\phi$ for $x > x_0$, then x_1 can be found so that $F > (H - \delta)\Phi$ for $x > x_1$: similarly $f < h\phi$ for $x > x_0$ involves $F < (h + \delta)\Phi$ for $x > x_1$.*

For if $f > H\phi$ for $x > x_0$, we have

$$F = \int_a^x f\, dt > \int_a^{x_0} f\, dt + H \int_{x_0}^x \phi\, dt > H\Phi + \int_a^{x_0} f\, dt - H \int_a^{x_0} \phi\, dt,$$

and if we choose x_1 so that

$$\left(\int_a^{x_0} f\, dt + H \int_a^{x_0} \phi\, dt \right) \Big/ \Phi < \epsilon$$

for $x \gtrless x_1$, as we certainly can if $\Phi \succ 1$, the result follows. Similarly in the other case. From this lemma we can at once deduce the following

Theorem. *Any one of the relations*

$$f \succ \phi, \, f \prec \phi, \, f \asymp \phi, \, f \asymp \phi, \, f \sim \phi$$

involves the corresponding one of the relations

$$F \succ \Phi, \, F \prec \Phi, \, F \asymp \Phi, \, F \asymp \Phi, \, F \sim \Phi$$

if either $F \succ 1$ *or* $\Phi \succ 1$.

To this we may add : *if both* $\int_x^\infty f \, dt$, $\int_x^\infty \phi \, dt$ *are convergent, then* $f \succ \phi, f \prec \phi, f \asymp \phi, f \asymp \phi, f \sim \phi$ *involve corresponding relations between*

$$\bar{F} = \int_x^\infty f \, dt, \quad \bar{\Phi} = \int_x^\infty \phi \, dt.$$

The proof we may leave to the reader. These results have been stated primarily for the case in which f and ϕ are positive ; but there is no difficulty in extending them to the case in which either function or both are negative.

2. Differentiation. It follows from § 1 that $f \succ \phi$ involves $f' \succ \phi'$ if $f \succ 1$ or $f \prec 1$ and *if any one of the relations expressed by* $\succ$, $\prec$, $\asymp$, $\asymp$, $\sim$ *holds between* f' *and* ϕ'.

In interpreting this statement regard must be paid to the conventions laid down in I. § 4. Thus if $f \succ \phi \succ 1$, f' and ϕ' are positive, and $f' \succ \phi'$. But if $f \succ 1 \succ \phi$, ϕ is a decreasing function and $\phi' < 0$. In this case $f' \succ -\phi'$, a relation which we have agreed to denote by $f' \succ \phi'$. If $1 \succ f \succ \phi$ both f' and ϕ' are negative: the relation $-f' \prec -\phi'$ would involve

$$-\int_x^\infty f' dt \prec -\int_x^\infty \phi' dt$$

or $f \prec \phi$, and is therefore impossible ; similarly for $-f' \asymp -\phi'$; so we must have $-f' \succ -\phi'$, a relation which we have agreed also to denote by $f' \succ \phi'$. The case in which $f \asymp 1$ is exceptional; any one of the relations $f' \succ \phi'$, etc. may then hold. Thus if $f = 1 + e^{-x}$, $f' = 1/x$, we have $f \succ \phi$, $f' \prec \phi'$. The fact is that in this case f, regarded as the integral of f', is dominated by the constant of integration.

Similar results hold, of course, for the relations $f \prec \phi$, etc., with similar exceptions. With regard to all of them it is to be observed that the assumption that one of the relations holds between f' and ϕ' is essential. We can never *infer* that one of them holds. We cannot even infer that f' or ϕ' is a steadily increasing or decreasing function at all. Thus if $f = e^x$, $\phi = e^x + \sin e^x$, we have $f' = e^x$ and $\phi' = e^x (1 + \cos e^x)$. Thus f and ϕ increase steadily and $f \sim \phi$, $f' \sim f$;

but ϕ' does not tend to infinity (vanishing for an infinity of values of x). Again if

$$\phi = e^x(\sqrt{2} + \sin x) + \tfrac{1}{2}x^2,$$

we have

$$\phi' = e^x(\sqrt{2} + \sin x + \cos x) + x$$

and $\phi \asymp e^x$, while ϕ' oscillates between the orders of e^x and x. It is possible, though less easy, to obtain examples of this character in which ϕ' also is monotonic.

3. Differentiation of L-functions. If f and ϕ are L-functions, so are f' and ϕ', and one of the relations $f' \succ \phi'$, $f' \asymp \phi'$, $f' \prec \phi'$ certainly holds (III. § 2). Thus in this case *both differentiation and integration are always legitimate**—this statement, however, being subject to certain exceptions in the cases in which $f \asymp 1$ or $\phi \asymp 1$.

In what follows we shall suppose that all the functions concerned are L-functions, or at any rate resemble L-functions in so far that one of the relations $f \succ \phi$, $f \asymp \phi$, $f \prec \phi$ is bound to hold between any pair of functions, and that differentiation and integration are permissible†.

1. *If f is an increasing function, and $f' \succ f$, then $f \succ e^{\Delta x}$. If $f' \prec f$, then $f \prec e^{\delta x}$. Similarly if f is a decreasing function, $f' \succ f$ and $f' \prec f$ involve $f \prec e^{-\Delta x}$ and $f \succ e^{-\delta x}$ respectively. If $f' \asymp f$, then $e^{\delta x} \prec f \prec e^{\Delta x}$ or $e^{-\Delta x} \prec f \prec e^{-\delta x}$, and we can find a number μ such that $f = e^{\mu x}f_1$, where $e^{-\delta x} \prec f_1 \prec e^{\delta x}$.*

The proofs of these assertions are almost obvious. Thus if f is an increasing function, and $f' \succ f$, we have

$$f'/f \succ 1, \quad \log f \succ x,$$

and so $\log f \succ \Delta x$ for $x \succ x_0$, *i.e.* $f \succ e^{\Delta x}$, or, what is the same thing, $f \succ e^{\Delta x}$. The last clause of the theorem follows at once from III. § 4.

2. *More generally, if v is any increasing function, $f'/f \succ v'/v$ involves $f \succ v^\Delta$ or $f \prec v^{-\Delta}$, according as f is an increasing or a decreasing function; and $f'/f \prec v'/v$ involves $f \prec v^\delta$ or $f \succ v^{-\delta}$. And $f'/f \asymp v'/v$ involves $v^\delta \prec f \prec v^\Delta$ or $v^{-\Delta} \prec f \prec v^{-\delta}$; and then we can find a number μ such that $f = v^\mu f_1$, where $v^{-\delta} \prec f_1 \prec v^\delta$.*

When f is an increasing function we shall call f'/f the *type t* of f‡: it being understood that t may be replaced by any simpler function τ such that $t \asymp \tau$. The type of a *decreasing* function f we define to be

---

* A tacit assumption to this effect underlies much of Du Bois-Reymond's work.

† The results which follow are all in substance due to Du Bois-Reymond.

‡ Du Bois-Reymond calls f/f' the type; the notation here adopted seems slightly more convenient.

the same as that of the increasing function $1/f$. The following table shews the types of some standard functions :

Function	1	llx	lx	x^a	e^x	e^{ax^β}	e_2x	e_3x	...
Type	0	$\dfrac{1}{x\,lx\,llx}$	$\dfrac{1}{x\,lx}$	$\dfrac{1}{x}$	1	$x^{\beta-1}$	ex	$e_2x\,ex$	...

If $f \succ \phi$, then $f'/f \succeq \phi'/\phi$. By making the increase of f large enough we can make the increase of $t = f'/f$ as large as we please. The reader will find it instructive to write out formal proofs of these propositions, and also of the following.

1. As the increase of f becomes smaller and smaller, f'/f tends to zero more and more rapidly, but, so long as $f \to \infty$ at all, we cannot have

$$f'/f \prec \phi, \quad \int^\infty \phi\,dx \text{ convergent.}$$

On the other hand, if the last integral is divergent we can find f so that $f \succ 1$, $f'/f \prec \phi$.

2. Although we can find f so that f'/f shall have an increase larger than that of any given function of x, we cannot have

$$f'/f \succ \phi(f), \quad \int^\infty \frac{dx}{x\phi(x)} \text{ convergent.}$$

On the other hand, if the last integral is divergent we can find f so that $f'/f \succ \phi(f)$.

[Thus we cannot find a function f which tends to infinity so slowly that $f'/f \prec 1/x^a$ $(a > 1)$. But we can find f so that $f'/f \prec 1/x\,lx\,llx$ (e.g. $f = l_3x$). We cannot find f so that $f'/f \succ f^a$ or $f' \succ f^{1+a}$ $(a > 0)$. But we can find f so that $f'/f \succ lf$ (e.g. $f = e_3x$).]

3. If $f \succ e_kx$ for all values of k, f'/f satisfies the same condition, and

$$f' \succ f\,lf\,l_2f\ldots l_kf.$$

He will also find it profitable to formulate corresponding theorems about functions of a positive variable x which tends to zero.

4. **Successive differentiation.** Du Bois-Reymond has given the following general theorem, which enables us to write down the increase of any derivative of any logarithmico-exponential function. We write t for f'/f, as in the last section, and we assume that no derivative $f^{(n)}$ satisfies $f^{(n)} \asymp 1$: if this should be the case the results of the theorem, so far as the derivatives $f^{(n+1)}, \ldots$ are concerned, cease to be true.

Theorem. (i) *If $t \succ 1/x$ (so that $f \succ x^\Delta$) then*

$$f \asymp f'/t \asymp f''/t^2 \asymp f'''/t^3 \ldots \asymp f^{(n)}/t^n \ldots.$$

(ii)   *If $t \prec 1/x$ (so that $f \prec x^\delta$) then*

$$f \asymp f'/t \asymp xf''/t \asymp x^2 f'''/t \ldots \asymp x^{n-1} f^{(n)}/t \ldots.$$

(iii)   *If $t \asymp 1/x$ (so that $f = x^\mu f_1$, where $x^{-\delta} \prec f_1 \prec x^\delta$), then if μ is not integral either set of formulae is valid. But if μ is integral*

$$f \asymp xf' \asymp x^2 f'' \ldots \asymp x^\mu f^{(\mu)} \asymp x^\mu f^{(\mu+1)}/t_1 \asymp x^{\mu+1} f^{(\mu+2)}/t_1 \ldots,$$

where t_1 is the type of f_1.

(i)   If $t \succ 1/x$, $1/t \prec x$ and so $t'/t^2 \prec 1$; hence $t'/t \prec t = f'/f$ or

$$ft' \prec f't.$$

Differentiating the relation $f' \asymp ft$, and using the relation just established, we obtain

$$f'' \asymp f't + ft' \asymp f't.$$

Thus the type of f' is the same as that of f; accordingly the argument may be repeated and the first part of the theorem follows.

(ii)   If $t \prec 1/x$, $xf' \prec f$ and so

$$xf'' + f' \prec f'.$$

But this cannot possibly be the case unless $xf'' \asymp f'$. Differentiating again we infer

$$xf''' + 2f'' \prec f'',$$

whence $xf''' \asymp f''$; and so on generally*. Thus the second part follows.

(iii)   If $t \asymp 1/x$, $f = x^\mu f_1$ and t_1, the type of f_1, satisfies $t_1 \prec 1/x$. Then

$$f' = \mu x^{\mu-1} f_1 + x^\mu f_1' \asymp x^{\mu-1} f_1 (\mu + xt_1) \asymp x^{\mu-1} f_1;$$

Similarly $f'' \asymp x^{\mu-2} f_1$ and so on. We can proceed indefinitely in this way unless μ is integral: in this case we find $f^{(\mu)} \asymp f_1$, and from this point we proceed as in case (ii).

Examples. (i) If $f = e^{\sqrt{x}}$, then $t = 1/\sqrt{x} \succ 1/x$, and $f^{(n)} \asymp e^{\sqrt{x}}/(\sqrt{x})^n$. If $f = e^{(\log x)^2}$, then $t = (\log x)/x \succ 1/x$, and $f^{(n)} \asymp e^{(\log x)^2} (\log x)^n/x^n$.

(ii)   If $f = (\log x)^m$, then $t = 1/(x \log x) \prec 1/x$, and

$$f^{(n)} \asymp tx^{-(n-1)} f \asymp (\log x)^{m-1}/x^n.$$

(iii)   If $f = x^2 llx$, $t \asymp 1/x$. Here

$$f' \asymp xllx, \quad f'' \asymp llx, \quad f''' \asymp 1/x lx, \quad f'''' \asymp 1/x^2 lx, \ldots.$$

(iv)   The results of the theorem, in the first two cases, can be stated more precisely as follows:

If $t \succ 1/x$, then

$$f^{(n)} \sim (f'/f)^n f.$$

---

* More precisely $xf'' \sim -f'$, $xf''' \sim -2f''$, and so on.

If $t < 1/x$, then
$$f^{(n)} \sim (-1)^{n-1} (n-1)! \, x^{-(n-1)} f'.$$

If f is a positive increasing function, then if $t > 1/x$ all the derivatives are ultimately positive, while if $t < 1/x$ they are alternately ultimately positive and ultimately negative.

5. Functions of an integral variable. The theorems for functions of an integral variable n, corresponding to those of §§ 1—4, involve sums
$$A_n = a_1 + a_2 + \ldots + a_n$$
in place of integrals, and differences
$$\Delta a_n = a_n - a_{n+1}$$
instead of differential coefficients. The reader will be able to formulate and to prove for himself the theorems which correspond to those of § 1. Thus

'$a_n > b_n$, $a_n < b_n$, $a_n \asymp b_n$, $a_n \not\asymp b_n$, $a_n \sim b_n$ involve the corresponding equations for A_n, B_n, if one at least of A_n, B_n tends to infinity with n'

and so on*. Considerations of space forbid that we should go further into the subject here.

VII.

SOME DEVELOPMENTS OF DU BOIS-REYMOND'S INFINITÄRCALCÜL.

1. WE shall conclude our account of the general theory by a brief sketch of some interesting results due in the main to Du Bois-Reymond. For further details we must refer to his memoirs catalogued in the Bibliographical Appendix.

$$\text{The functions } \frac{f(x+a)}{f(x)}, \ \frac{f(ax)}{f(x)}, \text{ etc.}$$

It is often necessary to obtain approximations to such functions as
$$f(x+a)/f(x),$$
where a is itself a function of x, which for simplicity we suppose positive, and which may tend to infinity with x. In this connection

---

* This is of course the well known theorem of Cauchy and Stolz: see Bromwich, *Infinite Series*, p. 377.

Du Bois-Reymond* has proved a whole series of theorems: it will be sufficient for our present purpose to give a few specimens of his results. In what follows it will be assumed throughout that all the functions dealt with are L-functions, or at any rate such that any pair of them satisfy one of the relations $f \succ \phi$, $f \asymp \phi$, $f \prec \phi$, and that such relations may be differentiated or integrated. This being so we have

$$\frac{f(x+a)}{f(x)} = e^{lf(x+a)-lf(x)} = e\left\{a\frac{f'(x+a)}{f(x+a)}\right\},$$

where $0 < a < a$. This expression has certainly the limit unity if $f' \preccurlyeq f$ and $a \prec 1$. Hence

$$f(x+a) \sim f(x) \qquad \ldots\ldots\ldots\ldots\ldots(1)$$

if $a \prec 1$ and $e^{-\Delta x} \prec f \prec e_{\Delta x}$. If $f'/f \prec 1$, i.e. if $e^{-\delta x} \prec f \prec e^{\delta x}$, the relation (1) holds for $a \prec f/f'$: it certainly holds, for instance, if $a = x\{f(x)\}^{-\mu}$, where $\mu > 0$, since $x/f^{\mu} \prec f/f'$ whenever $f \succ 1$†.

If $a \asymp f/f'$ (as e.g. if $f = e^{\mu x}f_1$, where $e^{-\delta x} \prec f_1 \prec e^{\delta x}$, and $a \asymp 1$), $f(x+a)/f(x)$ will tend to a limit different from unity.

Again $\qquad \dfrac{f(x+a)}{f(x)} = e\left\{a\dfrac{f'(x)}{f(x)}\dfrac{t(x+a)}{t(x)}\right\},$

where $t = f'/f$. Hence

$$\frac{f(x+a)}{f(x)} = e\left\{u\frac{f'(x)}{f(x)}\right\} \qquad (u \sim a)\ldots\ldots\ldots\ldots(2)$$

in all cases in which $t(x+a)/t(x) \sim 1$; as for example if $a \preccurlyeq 1$, $e^{-\delta x} \prec t \prec e^{\delta x}$, or, what is the same thing, if

$$a \preccurlyeq 1, \quad e^{-e^{\delta x}} \prec f \prec e^{e^{\delta x}}.$$

The reader will find it instructive to write down conditions under which the equation (2) holds when $u \asymp a$ is substituted for $u \sim a$, and to consider in what circumstances either relation holds when $a \succ 1$.

2. The reader is also recommended to verify some of the following results :

(i) *If* $1 \prec a \prec x$ *and* $x^{-\Delta} \prec f \prec x^{\Delta}$, *then* $f(x+a)/f(x) \sim 1$.

(ii) *If* $f \prec x$ *and* $a \prec 1/f'$, *or if* $f \asymp x$ *and* $a \prec 1$, *then* $f(x+a) - f(x) \prec 1$.

(iii) *If* $e^{-\delta x} \prec f \prec e^{\delta x}$ *and* $a \prec f'/f''$, *then*

$$f(x+a) - f(x) \sim af'(x).$$

---

* *Math. Annalen*, Bd. 8, S. 363 *et seq.*

† For $\displaystyle\int^{\infty} f^{-1-\mu}f' dx$ is convergent, and so $f'/f^{1+\mu} \prec 1/x$.

The condition $a \prec f'/f''$ may be simplified by means of the theorem of VI. § 4. Thus if $t \prec 1/x$ (i.e. if $f \prec x^\delta$) it is equivalent to $a \prec x$.

(iv) If $x^{-\delta} \prec a \prec x^\delta$, $(lx)^{-\Delta} \prec f \prec (lx)^\Delta$, then $f(ax)/f(x) \sim 1$.

(v) If $e^{-\Delta \sqrt{(lx)}} \prec f \prec e^{\Delta \sqrt{(lx)}}$, then

$$\frac{f\{xf(x)\}}{f(x)} \gtreqless 1, \quad e\left\{\frac{xlf(x) f'(x)}{f(x)}\right\} \gtreqless 1;$$

and the limits of the two functions are the same: and if $e^{-\delta \sqrt{(lx)}} \prec e^{\delta \sqrt{(lx)}}$ this limit is unity.

Suppose, e.g. $f \succ 1$, and let $f(x) = \phi(lx)$; then, if $a = f(x)$,

$$\frac{f(ax)}{f(x)} = e^{l\phi(lx+la)-l\phi(lx)} = e^{la\phi'(lx+la_1)/\phi(lx+la_1)},$$

where $1 < a_1 < a$. The exponent is

$$l\phi(lx+la_1) \frac{\phi'(lx+la_1)}{\phi(lx+la_1)} \frac{l\phi(lx)}{l\phi(lx+la_1)}.$$

Now $a = f(x) \prec x^\delta$ and therefore $la_1 \leqslant la \prec lx$, and so, by (i),

$$l\phi(lx+la_1) \sim l\phi(lx)$$

if $l\phi \prec x^\Delta$ or if $f \prec e^{(lx)^\Delta}$, which is certainly the case. Hence the exponent is asymptotically equivalent to

$$l\phi(u) \phi'(u)/\phi(u),$$

where $u = lx + la_1$. And $l\phi(\phi'/\phi) \leqslant 1$ if $(l\phi)^2 \leqslant u$, i.e. if $\phi \leqslant e^{\Delta \sqrt{u}}$ or $f \prec e^{\Delta \sqrt{(lx)}}$. In this case $f(ax) \gtreqless f(x)$; and it is easy to see that if $f \prec e^{\delta \sqrt{(lx)}}$ the symbol $\gtreqless$ may be replaced by $\sim$.

(vi) If $f(x) = x\phi(x)$, and $e^{-\delta \sqrt{(lx)}} \prec \phi \prec e^{\delta \sqrt{(lx)}}$, then

$$f_2(x) \equiv ff(x) \sim x\phi^2, \dots, f_n \sim x\phi^n, \dots.$$

The reader will easily prove this by the aid of the preceding results. He will also find it instructive to calculate the increase of f_n when $f = e^{\sqrt{(lx)}}$ and when $f = e^{(lx)^a}$, where $a > \frac{1}{2}$.

The accuracy of approximations.

3. We have already (IV. §§ 3—4) had occasion to use the notion of an approximation to the increase of a function, and to distinguish legitimate and illegitimate forms of approximation. Du Bois-Reymond has given the following more precise definitions.

He defines $\psi(x, u, u_1, \dots)$ to be an 'approximate form' of y if

$$y = \psi(x, u, u_1, \dots),$$

ψ being a known function, and $u, u_1, \dots$ unknown functions whose increase is, however, subject to certain limitations. It is clear that it is really useless, however, to insert more than one unknown function

u in ψ. The effect of the presence of u is to define a certain stretch within which the increase of y lies, and the presence of several u's can effect no more. We shall therefore consider only approximate forms of the type

$$y = \psi\,(x,\ u) \ \dots\dots\dots\dots\dots\dots\dots\dots(1).$$

Thus $e^{x^u}\ (u \sim 1),\ e^{(1+u)x}\ (u \prec 1),\ x^{1+u}e^x\ (u \prec 1)\dots\dots\dots(2)$

are approximate forms of $y = xe^x/lx$; the second clearly closer than the first and the third than the second.

The closeness of an approximation may be measured as follows. The presence of u in (1) lends a certain degree of indeterminateness to the increase of y: all that we can say (the increase of u being known to lie between certain limits) is that y lies in a certain interval

$$\eta_1 \preccurlyeq y \preccurlyeq \eta_2.$$

Now (II. § 8) we can find an increasing function F so that $F(\eta_1) \asymp F(\eta_2)$: if F satisfies this condition, any more slowly increasing function will do so too. *The slower the increase of F must be taken, the rougher the approximation.*

The facts may be stated the other way round. Given y, and a function F, such that $1 \prec F \prec x$, we can determine an interval $\eta_1 \preccurlyeq y \preccurlyeq \eta_2$ such that $F(\eta_1) \asymp F(\eta_2)$. The slower the increase of F, the larger this interval may be taken; if $F \asymp x$ it vanishes, if $F \asymp 1$ it may be taken as large as we please. If $F = lx$ it might be (y^δ, y^Δ); if $F = l_2x$ it might be

$$e^{(ly)^\delta},\ e^{(ly)^\Delta},$$

and so on. No logarithmico-exponential form of F, however, can give an interval as large as $(\log y,\ e^y)$; a function F such that $F(y) \asymp F(e^y)$ must transcend any logarithmico-exponential scale.

Let us consider the approximations (2) for xe^x/lx.

(i) If $y = e^{x^u}\ (u \sim 1)$, y lies in the interval $e^{x^{1-\delta}},\ e^{x^{1+\delta}}$. Since

$$ll\,(e^{x^{1-\delta}}) = (1 - \delta)\,lx \asymp ll\,(e^{x^{1+\delta}})$$

we may take $F = llx$, or even $F = (llx)^\Delta$: but the increase of F cannot be taken as large as $(lx)^\delta$.

(ii) If $y = e^{(1+u)x}\ (u \prec 1)$, y lies in the interval $e^{(1-\delta)x},\ e^{(1+\delta)x}$. Then we may take $F = (lx)^\Delta$, but we cannot take $F = e^{(lx)^\delta}$.

(iii) If $y = x^{1+u}e^x$ we may, as the reader will easily verify, take $F = e^{(lx)^\mu}$, where μ is any number less than unity.

Another example of an approximation is given by the formula

$$\frac{f(x+a)}{f(x)} = e\left\{u\frac{f'(x)}{f(x)}\right\} \quad (u \sim a).$$

If, *e.g.*, a is a constant,

$$l\left\{\frac{f(x+a)}{f(x)}\right\} \sim l\left\{e\left[\frac{f'(x)}{f(x)}\right]\right\},$$

and the degree of accuracy of the approximation is great enough to be measured by the function $F = lx$.

The approximate solution of equations.

4. It is often important to obtain an asymptotic solution of an equation $f(x, y) = 0$, *i.e.* to find a function whose increase gives an approximation to that of y. No very general methods of procedure can be given, but the kind of methods which may be pursued are worth illustrating by a few examples.

(i) Suppose that the equation is

$$x = y\kappa(y) \quad \dots\dots\dots\dots\dots\dots\dots(1),$$

where $y^{-\delta} \prec \kappa \prec y^{\delta}$. If the increase of κ is so slow that $\kappa\{y\kappa(y)\} \asymp \kappa(y)$ it is clear that

$$y \asymp x/\kappa(y) \asymp x/\kappa(x):$$

and if the increase of κ is slow enough we may have $y \sim x/\kappa(x)$.

The conditions

$$e^{-\Delta\sqrt{(ly)}} \prec \kappa(y) \prec e^{\Delta\sqrt{(ly)}}, \quad e^{-\delta\sqrt{(ly)}} \prec \kappa(y) \prec e^{\delta\sqrt{(ly)}}$$

are, by the result (v) of § 2, enough to ensure the truth of these hypotheses; and then $y = ux/\kappa(x)$, where $u \asymp 1$ (or $u \sim 1$) is an approximate solution of our equation.

Du Bois-Reymond has proved that the more elaborate approximations

$$y = ux/\{\kappa(x/\kappa)\}, \quad y = ux\kappa^{-1/\{1+(x\kappa'/\kappa)\}}$$

have a wider range of validity: and that more elaborate approximations still may be constructed valid within the range

$$e^{-\Delta(ly)^{1-\delta}} \prec \kappa \prec e^{\Delta(ly)^{1-\delta}}.$$

The more general equation

$$x = y^m\kappa(y)$$

can clearly be reduced to the form considered above by writing x^m for x and κ^m for κ.

In general, if $x = \phi(y)$, the more rapid the increase of ϕ the more precisely can we determine the increase of y as a function of x. Thus if

$$x = y e^y$$

we have $lx = y + ly$ and

$$y = lx - ly = lx\,(1 + u),$$

where $u \sim ly/lx \sim llx/lx$. This is a solution of a much more precise kind than those considered above.

5. The reader will find it instructive to examine the following results :

(i) Let $\qquad x = y e^{(ly)^{3/8}}$.

This is an example of the work of § 4: and

$$y \sim x e^{-(lx)^{3/8}}.$$

(ii) Let $\qquad x = y e^{(ly)^{5/8}}$.

Here $\qquad y \sim x e\left[-(lx)^{5/8}\{1 - (lx)^{-3/8}\}^{5/8} \right]$

$\qquad\qquad \sim x e\{ -(lx)^{5/8} + \tfrac{5}{8}(lx)^{1/4}\}$.

(iii) Let $\qquad x = y^m (ly)^{m_1} (l_2 y)^{m_2} \ldots (l_r y)^{m_r}$.

Here $\qquad y \sim m^{m_1/m} x^{1/m} (lx)^{-m_1/m} \ldots (l_r x)^{-m_r/m}$.

(iv) Let $\qquad x = e^{y^2} ly$.

Here $\qquad y = \sqrt{(lx - l_3 x)} + u \quad (u \!\prec\! 1)$.

(v) As an example of another type, Du Bois-Reymond has considered the equation

$$f(x+y) - f(x) = C,$$

where C is a positive constant. He finds

$$y \sim C/f'(x) \quad (f(x) \succ lx),$$
$$y = x e\{Cu/xf'(x)\} \quad (u \sim 1,\ lx \succ f(x) \succ llx),$$

and so on : the forms of the solution when $f \asymp lx$, $f \asymp llx$, ... are exceptional.

(vi) As an example of an approximation pushed to greater lengths let us take the following result: if

$$x = y\,ly,$$

then $\qquad y = \dfrac{x}{lx}\left\{ 1 + \dfrac{llx}{lx} + \dfrac{(llx)^2}{(lx)^2} - \dfrac{llx}{(lx)^2}\right\} + u,$

where $\qquad u \gtrless \dfrac{x\,(llx)^3}{(lx)^4}$.

6. Here we may bring our account of the general theory to a close. It is a theory that has found, and is finding, a large and increasing variety of applications in various branches of mathematics : the nature of some of these applications the reader may glean from Appendix II.

APPENDIX I.

GENERAL BIBLIOGRAPHY.

DU BOIS-REYMOND'S memoirs bearing on the subjects of this tract are :
Sur la grandeur relative des infinis des fonctions (*Annali di Matematica*, Serie 2, t. 4, p. 338).

Théorème général concernant la grandeur relative des infinis des fonctions et de leurs derivées (*Crelle's Journal*, Bd. 74, S. 294).

Eine neue Theorie der Convergenz und Divergenz von Reihen mit positiven Gliedern. *Anhang:* Ueber die Tragweite der logarithmischen Kriterien (*Crelle's Journal*, Bd. 76, S. 61).

Ueber asymptotische Werthe, infinitäre Approximationen, und infinitäre Auflösung von Gleichungen (*Math. Annalen*, Bd. 8, S. 363). Nachtrag zur vorstehenden Abhandlung (*ibid.*, S. 574).

Notiz über infinitäre Gleichheiten (*Math. Annalen*, Bd. 10, S. 576).

Ueber die Paradoxen des Infinitärcalcüls (*Math. Annalen*, Bd. 11, S. 149).

Notiz über Convergenz von Integralen mit nicht verschwindendem Argument (*Math. Annalen*, Bd. 13, S. 251).

Ueber Integration und Differentiation infinitären Relationen (*Math. Annalen*, Bd. 14, S. 498).

Ueber den Satz : $\lim f'(x) = \lim f(x)/x$ (*Math. Annalen*, Bd. 16, S. 550).

See also

A. PRINGSHEIM: Ueber die sogenannte Grenze und die Grenzgebiete zwischen Convergenz und Divergenz (*Münchener Sitzungsberichte*, Bd. 26, S. 605).

—— Ueber die Du Bois-Reymond'sche Convergenz-Grenze u.s.w. (*Münchener Sitzungsberichte*, Bd. 27, S. 303).

—— Allgemeine Theorie der Convergenz und Divergenz von Reihen mit positiven Gliedern (*Math. Annalen*, Bd. 35, S. 347).

—— Zur Theorie der bestimmten Integrale und der unendlichen Reihen (*Math. Annalen*, Bd. 37, S. 591).

J. HADAMARD: Sur les caractères de convergence des séries à termes positifs et sur les fonctions indéfiniment croissantes (*Acta Mathematica*, t. 18, p. 319 and p. 421).

S. PINCHERLE : Alcune osservazioni sugli ordini d' infinito delle funzioni (*Memorie della Accademia delle Scienze di Bologna*, Ser. 4, t. 5, p. 739).

E. BOREL : *Leçons sur les fonctions entières*, pp. 111—122.

—— *Leçons sur les séries à termes positifs*, pp. 1—50.

—— *Leçons sur la théorie de la croissance*.

APPENDIX II.

A SKETCH OF SOME APPLICATIONS*, WITH REFERENCES.

A. *Convergence and divergence of series and integrals.*

(i) *The logarithmic tests.* The series $\Sigma u_n \, (u_n \geqslant 0)$ is convergent if

$$u_n \preccurlyeq n^{-1-a}$$

$$\text{or} \qquad u_n \preccurlyeq (n \ln \dots l_{k-1} n)^{-1} (l_k n)^{-1-a},$$

where $a > 0$, and divergent if

$$u_n \succ n^{-1}$$

$$\text{or} \qquad u_n \succcurlyeq (n \ln \dots l_k n)^{-1}.$$

The integral $\displaystyle\int^{\infty} f(x) \, dx \ (f \geqq 0)$ is convergent if

$$f \preccurlyeq x^{-1-a}$$

$$\text{or} \qquad f \preccurlyeq (x \, lx \dots l_{k-1} x)^{-1} (l_k x)^{-1-a},$$

where $a > 0$, and divergent if

$$f \preccurlyeq x^{-1}$$

$$\text{or} \qquad f \succcurlyeq (x \, lx \dots l_k x)^{-1}.$$

The integral $\displaystyle\int_0 f(x) \, dx \, (f \geqq 0)$ is convergent if

$$f \preccurlyeq (1/x)^{1-a}$$

$$\text{or} \qquad f \preccurlyeq (1/x) \{l (1/x) \dots l_{k-1} (1/x)\}^{-1} \{l_k (1/x)\}^{-1-a},$$

where $a > 0$, and divergent if

$$f \succ 1/x$$

$$\text{or} \qquad f \succcurlyeq (1/x) \{l (1/x) \dots l_k (1/x)\}^{-1}.$$

* That is to say of certain regions of mathematical theory in which the notation and the ideas of the *Infinitärcalcül* may be used systematically with a great gain in clearness and simplicity.

[The first general statement of the 'logarithmic criteria,' so far as series are concerned, appears to have been made by De Morgan : see his *Differential and Integral Calculus,* 1839, p. 326. The essentials of the matter, however, appear in a posthumous memoir of Abel (*Œuvres complètes,* t. 2, p. 200 ; see also t. 1, p. 399). This memoir appears also to have been first published in 1839. The case of $k=1$ had been dealt with by Cauchy (*Exercices de Mathématiques,* t. 2, 1827, pp. 221 *et seq.*). Bertrand appears to have arrived at some or all of De Morgan's results independently (see *Liouville's Journal,* t. 7, 1842, p. 37) and the criteria are very commonly attributed to him. The criteria for integrals do not appear to have been stated generally before Riemann, *Inaugural-Dissertation* of 1854 (*Werke,* S. 229).

The following references may also be useful :

Bonnet, *Liouville's Journal,* t. 8, p. 78.

Dini, *Sulle serie a termini positivi* (Pisa, 1867); also in the *Annali dell' Univ. Tosc.,* t. 9, p. 41.

Du Bois-Reymond, *Crelle's Journal,* Bd. 76, S. 619.

Pringsheim, *Math. Annalen,* Bd. 35, S. 347 and Bd. 37, S. 591 ; also in the *Encyklopädie der Math. Wiss.,* Bd. 1, Th. 1, S. 77 *et seq.*

Hobson, *Theory of functions of a real variable,* p. 406.

Bromwich, *Infinite series,* pp. 29, 37.

Hardy, *Course of pure mathematics,* pp. 357 *et seq.*

Chrystal, *Algebra,* vol. 2, pp. 109 *et seq.*]

(ii) *General theorems analogous to Du Bois-Reymond's Theorem* (II. § 1).

Given any divergent series Σu_n of positive terms, we can find a function v_n such that $v_n \prec u_n$ and Σv_n is divergent; *i.e.* given any divergent series we can find one more slowly divergent.

Given any convergent series Σu_n of positive terms we can find v_n so that $v_n \succ u_n$ and Σv_n is convergent; *i.e.* given any convergent series we can find one more slowly convergent.

Given any function $\phi(n)$ tending to infinity, however slowly, we can find a convergent series Σu_n and a divergent series Σv_n such that $v_n/u_n = \phi(n)$.

Given an infinite sequence of series, each converging (diverging) more slowly than its predecessor, we can find a series which converges (diverges) more slowly than any of them.

[See Abel and Dini, *l.c. supra* ; Hadamard, *Acta Mathematica,* t. 18, p. 319 and t. 27, p. 177; Bromwich, *Infinite series,* p. 40; Littlewood, *Messenger of Mathematics,* vol. 39, p. 191.]

H.   4

There is no function $\phi(n)$ such that $u_n\phi(n) \succ 1$ is a necessary condition for the divergence of Σu_n, and no function $\phi(n)$ such that $\phi(n) \succ 1$ and $u_n\phi(n) \preccurlyeq 1$ is a necessary condition for the convergence of Σu_n.

If u_n is a *steadily decreasing* function of n, then $nu_n \prec 1$ *is* a necessary condition for convergence; but there is no function $\phi(n)$ such that $\phi(n) \succ 1$ and $n\phi(n)u_n \prec 1$ is a necessary condition. [Pringsheim, *Math. Annalen*, Bd. 35, S. 343 *et seq.*; *ibid.*, Bd. 37, S. 591 *et seq.*]

If however nu_n decreases steadily, then $n \log nu_n \to 0$ is a necessary condition; and if $n\psi(n)u_n$, where $n\psi(n) \succ 1$ and $\int \dfrac{dn}{n\psi(n)} \succ 1$, decreases steadily, then

$$\left(n\psi(n) \int \frac{dn}{n\psi(n)} \right) u_n \to 0$$

is a necessary condition.

(iii) *Special series and integrals possessing peculiarities in respect to the mode of their convergence or divergence.*

For examples of series and integrals which converge or diverge so slowly as not to answer to any of the logarithmic criteria see Du Bois-Reymond, Pringsheim, Borel (*l.c. supra*), and Blumenthal, *Principes de la théorie des fonctions entières d'ordre infini*, ch. 1.

In these cases the logarithmic tests are insufficient (*en défaut*, IV. §§ 2, 5). For examples of series and integrals to which the logarithmic tests are *inapplicable* (v. §§ 3, 4) see the writings just mentioned and also

Thomae : *Zeitschrift für Mathematik*, Bd. 23, S. 68.

Gilbert : *Bulletin des Sciences Mathématiques*, t. 12, p. 66.

Goursat : *Cours d'Analyse*, t. 1, p. 205.

Hardy : *Messenger of Mathematics*, vol. 31, p. 1 ; *ibid.*, vol. 31, p. 177 ; *ibid.*, vol. 39, p. 28.

B. *Asymptotic formulae for finite series and integrals.*

A closely connected problem is that of the determination of asymptotic formulae for

$$A_n = a_1 + a_2 + \ldots + a_n$$

or for

$$\Phi(x) = \int_a^x \phi(t)\, dt,$$

when the behaviour of a_n or $\phi(x)$ for large values of n or x is known. A good deal can be accomplished in this direction by means of

(i) the theorem of Cauchy and Stolz, that, if a_n and b_n are positive and $a_n \sim Cb_n$, then $A_n \sim CB_n$, (ii) the theorems of VI. and (iii) the theorem of Maclaurin and Cauchy, that

$$\phi(1) + \phi(2) + \ldots + \phi(n) - \int_1^n \phi(x)dx,$$

where $\phi(x)$ is a positive and decreasing function of x, tends to a limit as $n \to \infty$.

[For (i) see Cauchy, *Analyse algébrique*, p. 52; Stolz, *Math. Annalen*, Bd. 14, S. 232, or *Allgemeine Arithmetik*, Bd. 1, S. 173; Jensen, *Tidskrift for Mathematik* (5), Bd. 2, S. 81; Bromwich, *Infinite series*, p. 378, and *Proc. Lond. Math. Soc.*, ser. 2, vol. 7, p. 101. Proofs of (iii) will be found in almost any modern treatise on analysis : *e.g.*, Bromwich, *Infinite series*, p. 29; Hardy, *Course of pure mathematics*, p. 305. An important extension to *slowly oscillating* series has been given recently by Bromwich (*Proc. Lond. Math. Soc.*, ser. 2, vol. 7, p. 327).]

Among the most important results which follow from these theorems are

$$1^s + 2^s + \ldots + n^s \sim \frac{n^{s+1}}{s+1} \quad (s > -1),$$

$$1^s + 2^s + \ldots + n^s - \frac{n^{s+1}}{s+1} \sim \zeta(-s) \quad (-1 < s < 0),$$

$$1 + \frac{1}{2} + \ldots + \frac{1}{n} - \log n \sim \gamma,$$

$$1 + \frac{a \cdot \beta}{1 \cdot \gamma} + \frac{a(a+1)\beta(\beta+1)}{1 \cdot 2 \cdot \gamma(\gamma+1)} + \ldots \text{ to } n \text{ terms},$$

$$\sim \frac{\Gamma(\gamma)}{\Gamma(a)\Gamma(\beta)} \frac{n^{a+\beta-\gamma}}{a+\beta-\gamma} \quad (a+\beta > \gamma),$$

$$or \qquad \sim \frac{\Gamma(a+\beta)}{\Gamma(a)\Gamma(\beta)} \log n \quad (a+\beta = \gamma).$$

In connection with the last result see Bromwich, *Proc. Lond. Math. Soc.*, ser. 2, vol. 7, p. 101; in the earlier formulae γ is Euler's constant and ζ denotes the 'Riemann ζ-function.'

The most important of all formulae of this kind is beyond question

$$\log 1 + \log 2 + \ldots + \log n - (n + \tfrac{1}{2})\log n + n \sim \tfrac{1}{2}\log(2\pi),$$

which, in the form

$$n! \sim n^{n+\frac{1}{2}}e^{-n}\sqrt{(2\pi)},$$

constitutes *Stirling's Theorem*. The literature connected with Stirling's Theorem and its extensions to the Gamma-function of a non-integral or complex variable is far too extensive to be summarized here. See *Encykl. der Math. Wiss.*, Bd. II. (2), S. 165 *et seq.* ; Bromwich, *Infinite series*, pp. 461 *et seq.*

Another formula of the same kind is

$$1^1 2^2 3^3 \ldots n^n \sim A n^{\frac{1}{2}n^2 + \frac{1}{2}n + \frac{1}{12}} e^{-\frac{1}{4}n^2},$$

where A is a constant defined by the equation

$$\log A = \tfrac{1}{12} \log 2\pi + \tfrac{1}{12}\gamma + \frac{1}{2\pi^2} \sum_1^\infty \frac{\log \nu}{\nu^2}.$$

The properties of this constant have been investigated by Kinkelin and Glaisher (Kinkelin, *Crelle's Journal*, Bd. 57, S. 122 : Glaisher, *Messenger of Mathematics*, vol. 6, p. 71 ; vol. 7, p. 43 ; vol. 23, p. 145 ; vol. 24, p. 1 ; *Quarterly Journal of Mathematics*, vol. 26, p. 1 : see also Barnes, *ibid.*, vol. 31, pp. 264 *et seq.*).

All these results are intimately bound up with the theory of the general 'Euler-Maclaurin Sum Formula'

$$\sum_1^n f(n) = \int^n f(x)\, dx + C + \tfrac{1}{2} f(n) + \frac{B_1}{2!} f'(n) - \frac{B_2}{4!} f'''(n) + \ldots$$

which also possesses an extensive literature (see Schlömilch, *Theorie der Differenzen und Summen*; Boole, *Finite differences*; Markoff, *Differenzenrechnung* ; Seliwanoff, *Differenzenrechnung* ; *Encykl. der Math. Wiss.*, Bd. I. S. 929 *et seq.* ; Bromwich, *Infinite series*, p. 238 and p. 324 ; Barnes, *Proc. Lond. Math. Soc.*, ser. 2, vol. 3, pp. 253 *et seq.* ; where many further references are given).

A simple example of the use of the general formula is afforded by the relation

$$\sum_1^n \nu^s - \frac{n^{s+1}}{s+1} - \tfrac{1}{2} n^s - \sum_1 (-1)^{i-1} \binom{s}{2i-1} \frac{B_i}{2i} n^{s-2i+1} \sim \zeta(-s).$$

Here s is positive and not integral, and the summation with respect to i is continued until we come to a negative power of n.

C. *Formulae involving prime numbers only.*

Asymptotic formulae involving functions defined arithmetically, and particularly functions defined by sums of functions of prime numbers only, play a most important part in the analytical theory of numbers. Of these the most important is the formula

$$\Pi(n) \sim \frac{n}{ln},$$

where $\Pi(n)$ denotes the number of prime numbers less than n.

Similarly it is known that

$$\Sigma lp \sim n, \ \Sigma \frac{lp}{p} \sim ln, \ \Sigma \frac{1}{p} \sim lln$$

(the summation in each case applying to all primes less than n) while $\overset{\infty}{\Sigma} \frac{1}{plp}$ is convergent.

Many more accurate results have been established by recent writers, particularly Mertens, Hadamard, Von Mangoldt, De la Vallée-Poussin, and Landau ; and the theory has to a considerable extent been freed from Riemann's still unproved assumption that all the roots of his Zeta-function have their real part equal to $\frac{1}{2}$. Thus it has been shown that

$$\Pi(n) = \int_2^n \frac{dx}{\log x} + O\left\{\frac{n}{(ln)^\Delta}\right\},$$

or, still more accurately,

$$\Pi(n) = \int_2^n \frac{dx}{\log x} + O\{ne^{-a\sqrt{(ln)}}\},$$

where a is a positive constant; but it still remains to be settled whether (as there is some reason to suppose) the last term can be replaced by $O(\sqrt{n})$ or even by

$$O\left(\frac{\sqrt{n}}{ln}\right).$$

[It would carry us too far to give detailed references to the literature of this exceedingly difficult and fascinating subject. The reader should consult Landau's exhaustive *Handbuch der Lehre von der Verteilung der Primzahlen* (Teubner, 1909).]

D.　*The theory of integral functions.*

1. The series $\Sigma c_n x^n$ will converge for all values of x (real or complex), and so define an *integral function* $f(x)$, if and only if $\sqrt[n]{|c_n|} \to 0$, i.e. if $|c_n| \prec e^{-\Delta n}$.

2. *The three indices of a function of finite order.* The three most important characters of an integral function $f(x)$ are :

(i)　$\gamma_n = |c_n|$, the modulus of the nth coefficient ;

(ii)　$a_n = |a_n|$, the modulus of the nth (in order of absolute magnitude) zero of $f(x)$;

(iii)　$M(r)$, the maximum of $|f(x)|$ on the circle $|x| = r$. $M(r)$ is known to be an increasing function of r, and in all cases $M(r) \succ r^\Delta$.

A function such that $M(r) \prec e^{r^\Delta}$ is called a *function of finite order*. We shall confine our remarks to such functions.

The principal problem of the theory of integral functions is the determination of the relations between the increases of a_n, $1/\gamma_n$, and $M(r)$. Those which subsist between the two latter functions are the simplest: when a_n is taken into account the theory is complicated by the 'Picard case of exception'—the case of functions which (like e^x) have no zeroes, or whose zeroes are scattered abnormally widely over the plane.

The nature of the results of the general theory may be gathered from a statement of a few of the simplest of them.

If
$$n^{-\mu-\delta} \prec \sqrt[n]{\gamma_n} \prec n^{-\mu+\delta},$$

i.e. if
$$l(1/\gamma_n) \sim \mu n \, ln,$$

we call μ the *μ-index*. The index may be defined in *all* cases without any assumption as to the existence of a limit for $\{l(1/\gamma_n)/(n\,ln)\}$; we confine ourselves to the simplest case.

If
$$n^{(1/\lambda)-\delta} \prec a_n \prec n^{(1/\lambda)+\delta},$$

we call λ the *λ-index*; and if
$$e^{r^{\nu-\delta}} \prec M(r) \prec e^{r^{\nu+\delta}},$$

we call ν the *ν-index*: thus
$$la_n \sim (ln)/\lambda, \quad ll\,M(r) \sim \nu\,lr.$$

Then $\mu = 1/\nu$: and *in general* $\lambda = \nu$.

Thus for the function
$$\frac{\sin(\sqrt{x})}{\sqrt{x}} = 1 - \frac{x}{3!} + \frac{x^2}{5!} - \dots$$

we have $\lambda = \nu = \frac{1}{2}$ and $\mu = 2$, as the reader will easily verify (using Stirling's Theorem to determine μ).

3. *Special results.* More precise results than these have been obtained in many cases. Thus if
$$\{n(ln)^{-a_1} \dots (l_\nu n)^{-a_\nu+\delta}\}^{-1/\rho} \prec \sqrt[n]{\gamma_n} \prec \{n(ln)^{-a_1} \dots (l_\nu n)^{-a_\nu-\delta}\}^{-1/\rho},$$

then
$$e\{r^\rho (lr)^{a_1} \dots (l_\nu r)^{a_\nu-\delta}\} \prec M(r) \prec e\{r^\rho (lr)^{a_1} \dots (l_\nu r)^{a_\nu+\delta}\},$$

and conversely.

As examples of still more accurate, but more special results, we may quote the following :

$$\Sigma \frac{x^n}{n^{an}} \sim \sqrt{\left(\frac{2\pi}{ea}\right)} \, x^{1/2a} \, e^{(a/e)x^{1/a}},$$

$$\Sigma \frac{x^n}{(n!)^a} \sim \frac{1}{\sqrt{a}} \, (2\pi)^{(1-a)/2} \, x^{(1-a)/2a} \, e^{ax^{1/a}},$$

$$\Sigma \frac{x^n}{\Gamma(an+1)} \sim (1/a) \, e^{x^{1/a}},$$

$$\Sigma e^{-n^p} x^n \sim \sqrt{\left\{\frac{2\pi}{p(p-1)}\right\}} \, \left(\frac{\log x}{p}\right)^{\frac{2-p}{2p-2}} \, e^{(p-1)\left(\frac{\log x}{p}\right)^{p/(p-1)}},$$

where $a > 0$ and in the last formula $1 < p < 2$, and throughout $x \to \infty$ by positive values.

These results may of course be used to give an upper limit for the modulus of the particular function considered when x is not necessarily real, and so for $M(r)$. Thus in the first case

$$M(r) \ll r^{1/2a} \, e^{(a/e)x^{1/a}}.$$

[The reader who wishes to become familiar with the theory of integral functions should begin by reading Borel's *Leçons sur les fonctions entières*. Some additions will be found in the notes at the end of the same writer's *Leçons sur les fonctions méromorphes*. He should then read two memoirs by E. Lindelöf; a short one in the *Bulletin des Sciences Mathématiques*, t. 27, p. 1, and a long one in the *Acta Societatis Fennicae*, t. 31, p. 1. Some of the results of this last paper were proved independently by Boutroux (*Acta Mathematica*, t. 28, pp. 97 *et seq.*); but M. Boutroux's important memoir is largely occupied by a discussion of some of the most difficult points in the theory.

Much of the theory has been developed in a very simple and elementary way by Pringsheim (*Math. Annalen*, Bd. 58, S. 257); and the reader should certainly consult a short note by Le Roy (*Bulletin des Sciences Mathématiques*, t. 24, p. 245). But, after reading the works of Borel and Lindelöf mentioned above, he will be wise to turn to Vivanti's *Teoria delle funzioni analitiche* (German translation by Gutzmer), which contains by far the fullest treatment of the subject yet published, and an exhaustive list of original memoirs.]

E. *Power series with a finite radius of convergence.*

Suppose that $a_1 + a_2 + \ldots$ is a divergent series: for simplicity suppose that a_n is always positive and steadily increases or decreases as n increases. Further suppose $e^{-\delta n} \prec a_n \prec e^{\delta n}$, so that $\Sigma a_n x^n$ is convergent if $0 \le x < 1$. Then a large number of interesting results have been established connecting the increase of a_n, as $n \to \infty$, and that of $f(x) = \Sigma a_n x^n$ as $x \to 1$. The fundamental result is: *if* $a_n \sim Cb_n$, *or, more generally, if* $(a_1 + a_2 + \ldots + a_n) \sim C(b_1 + b_2 + \ldots + b_n)$, *and* $f(x) = \Sigma a_n x^n$, $g(x) = \Sigma b_n x^n$, *then*

$$f(x) \sim Cg(x).$$

From this theorem it may be deduced that

$$\Sigma \frac{x^n}{n^p} \sim \frac{\Gamma(1-p)}{(1-x)^{1-p}} \qquad (p < 1),$$

$$F(a, \beta, \gamma, x) \sim \frac{\Gamma(\gamma)\,\Gamma(a + \beta - \gamma)}{\Gamma(a)\,\Gamma(\beta)} \frac{1}{(1-x)^{a+\beta-\gamma}} \qquad (a + \beta > \gamma)$$

$$F(a, \beta, a + \beta, x) \sim \frac{\Gamma(a+\beta)}{\Gamma(a)\,\Gamma(\beta)} l\left(\frac{1}{1-x}\right).$$

Of further results the following is typical: if

$$a_n \sim n^p / n \, l n \, \ldots \, l_{m-1} n \, (l_m n)^q \ldots (l_{m+k} n)^{q_k},$$

then

$$F(x) \sim \Gamma(p) \Big/ \Big\{ (1-x)^{p+1}$$

$$\times \frac{1}{1-x} l \frac{1}{1-x} \ldots l_{m-1} \frac{1}{1-x} \left(l_m \frac{1}{1-x}\right)^q \ldots \left(l_{m+k} \frac{1}{1-x}\right)^{q_k} \Big\}$$

if $p > 0$, $q \neq 1$: but

$$F(x) \sim 1 \Big/ \Big\{ (1-q)\left(l_m \frac{1}{1-x}\right)^{q-1} \left(l_{m+1} \frac{1}{1-x}\right)^{q_1} \ldots \left(l_{m+k} \frac{1}{1-x}\right)^{q_k} \Big\}$$

if $p = 0$, $q < 1$ (if $p < 0$ or $p = 0$, $q > 1$, then Σa_n is convergent).

Thus, *e.g.*

$$\Sigma \frac{n^p x^n}{(lx)^q} \sim \Gamma(p+1) \Big/ \Big\{ (1-x)^{p+1} \left(l \frac{1}{1-x}\right)^q \Big\}.$$

As specimens of further results of this character we may quote

$$x + x^4 + x^9 + \ldots \sim \tfrac{1}{2}\sqrt{\left(\frac{\pi}{1-x}\right)},$$

$$x + x^a + x^{a^2} + \ldots \sim \frac{1}{la}\, l\left(\frac{1}{1-x}\right) \qquad (a > 1),$$

$$\Sigma a^n x^{n^2} \sim e\left\{\tfrac{1}{4}\frac{(la)^2}{l(1/x)}\right\} \qquad (a > 1),$$

$$\Sigma e^{n/ln}\, x^n = e_2\{u/(1-x)\} \qquad (u \sim 1).$$

Many similar results have been established about series other than power series: thus

$$\Sigma \frac{x^n}{n(1+x^n)} \sim \tfrac{1}{2}\, l\left(\frac{1}{1-x}\right),$$

$$\Sigma \frac{x^n}{1-x^n} \sim \frac{1}{1-x}\, l\left(\frac{1}{1-x}\right).$$

As an example of a more precise result we may quote the formula

$$\Sigma \frac{x^n}{1+x^{2n}} = \tfrac{1}{4}\left\{\frac{\pi}{l(1/x)} - 1\right\} + O\{(1-x)^\Delta\}.$$

[See

Bromwich, *Infinite series*, pp. 131 *et seq.*, 171 *et seq.* ;

Le Roy, *Bulletin des Sciences Mathématiques*, t. 24, pp. 245 *et seq.* ;

Lasker, *Phil. Trans. Roy. Soc.*, (A), vol. 196, p. 433 ;

Pringsheim, *Acta Mathematica*, t. 28, p. 1 ;

Barnes, *Proc. Lond. Math. Soc.*, vol. 4, p. 284 ; *Quarterly Journal*, vol. 37, p. 289 ;

Hardy, *Proc. Lond. Math. Soc.*, vol. 3, p. 381 ; *ibid.*, vol. 5, p. 197 ; *ibid.*, vol. 5, p. 342 ;

where further references will be found. These writers also consider the extensions of such results to the field of the complex variable.]

APPENDIX III.

SOME NUMERICAL ILLUSTRATIONS.

Mr J. Jackson, scholar of Trinity College, has been kind enough to calculate for me the following numerical results, which will, I think, be found instructive as comments on some of the matters dealt with in the body of this tract and in Appendix II. It will of course be understood that, except in one or two instances, they are approximations and sometimes quite rough approximations.

1. *Table of the functions* $\log x$, $\log \log x$, $\log \log \log x$, *etc.*

x	$\log x$	$\log_2 x$	$\log_3 x$	$\log_4 x$	$\log_5 x$
10	$2\cdot30$	$0\cdot834$	$-0\cdot182$	——	——
10^3	$6\cdot91$	$1\cdot933$	$0\cdot659$	$-0\cdot417$	——
10^6	$13\cdot82$	$2\cdot626$	$0\cdot966$	$-0\cdot035$	——
10^{10}	$23\cdot03$	$3\cdot137$	$1\cdot143$	$0\cdot134$	$-2\cdot011$
10^{15}	$34\cdot54$	$3\cdot542$	$1\cdot265$	$0\cdot235$	$-1\cdot449$
10^{20}	$46\cdot05$	$3\cdot830$	$1\cdot343$	$0\cdot295$	$-1\cdot221$
10^{30}	$69\cdot08$	$4\cdot235$	$1\cdot443$	$0\cdot367$	$-1\cdot003$
10^{60}	$138\cdot15$	$4\cdot928$	$1\cdot595$	$0\cdot467$	$-0\cdot762$
10^{100}	$230\cdot26$	$5\cdot439$	$1\cdot693$	$0\cdot527$	$-0\cdot641$
10^{1000}	$2302\cdot58$	$7\cdot742$	$2\cdot047$	$0\cdot716$	$-0\cdot334$
10^{10^6}	2303×10^3	$14\cdot650$	$2\cdot685$	$0\cdot987$	$-0\cdot013$
$10^{10^{10}}$	2303×10^7	$23\cdot860$	$3\cdot172$	$1\cdot154$	$0\cdot144$

2. Table of the functions e^x, e^{e^x}, $e^{e^{e^x}}$, etc.

x	ex	e_2x	e_3x	e_4x
1	2·718	15·154	3,814,260	$10^{1,656,510}$
2	7·389	1618·2	$5·85 \times 10^{702}$	—
3	20·085	$5·28 \times 10^8$	$10^{2·295 \times 10^8}$	—
5	148·413	$2·85 \times 10^{64}$	$10^{1·24 \times 10^{64}}$	—
10	22026	$9·44 \times 10^{9505}$	—	—

The function $\log x$ is defined only for $x > 0$, $\log_2 x$ for $x > 1$, $\log_3 x$ for $x > e$, $\log_4 x$ for $x > e^e = e_2$, and so on. The values of the first few numbers e, e_2, e_3, ... are given above, viz. $e = 2·718$, $e_2 = 15·154$, $e_3 = 3,814,260$, $e_4 = 10^{1,656,510}$.

3. Table of the functions $n!$, n^n, n^{n^n}.

n	$n!$	n^n	n^{n^n}
1	1	1	1
2	2	4	16
3	6	27	$7·634 \times 10^{12}$
4	24	256	$1·491 \times 10^{154}$
5	120	3,125	$9·55 \times 10^{2,184}$
6	720	46,656	$2·7 \times 10^{36,305}$
7	5,040	823,543	$1·4 \times 10^{695,974}$
8	40,320	16,827,216	$10^{15,151,345}$
9	362,880	$3·8742 \times 10^8$	$10^{369,693,100}$
10	3,628,800	10^{10}	$10^{10,000,000,000}$
100	$9·346 \times 10^{157}$	10^{200}	—
10^{10}	$10^{9·57 \times 10^{10}}$	$10^{10^{11}}$	—

4. Table to illustrate the convergence of the series

$$(1)\ \sum_{3}^{\infty} \frac{1}{n \log n (\log \log n)^2}. \qquad (2)\ \sum_{2}^{\infty} \frac{1}{n (\log n)^2}. \qquad (3)\ \sum_{1}^{\infty} \frac{1}{n^s}\ (s = 1\text{·}1).$$

$$(4)\ \sum_{1}^{\infty} \frac{1}{n^s}\ (s = 1\text{·}5). \qquad (5)\ \sum_{1}^{\infty} \frac{1}{n^s}\ (s = 2). \qquad (6)\ \sum_{1}^{\infty} \frac{1}{n^s}\ (s = 10).$$

$$(7)\ \sum_{1}^{\infty} \frac{1}{n^s}\ (s = 100). \qquad (8)\ \sum_{0}^{\infty} x^n\ (x = \text{·}9). \qquad (9)\ \sum_{0}^{\infty} x^n\ (x = \text{·}5).$$

$$(10)\ \sum_{0}^{\infty} x^n\ (x = \text{·}1). \qquad (11)\ 1 + \frac{1}{2!} + \frac{1}{3!} + \dots \qquad (12)\ 1 + \frac{1}{2^2} + \frac{1}{3^3} + \dots$$

$$(13)\ \sum_{0}^{\infty} x^{n^2}\ (x = \text{·}9). \qquad (14)\ \sum_{0}^{\infty} x^{n^2}\ (x = \text{·}5). \qquad (15)\ \sum_{0}^{\infty} x^{n^2}\ (x = \text{·}1).$$

$$(16)\ \frac{1}{1^{1^1}} + \frac{1}{2^{2^2}} + \frac{1}{3^{3^3}} + \dots$$

Series	Sum	Number of terms required to calculate the sum correctly to			
		2	10	100	1000
			decimal places *		
1	38·43	$10^{3\text{·}14 \times 10^{86}}$	—	—	—
2	2·11	$7\text{·}23 \times 10^{86}$	$10^{8\text{·}6 \times 10^9}$	—	—
3	10·58	10^{33}	10^{113}	10^{1013}	10^{10013}
4	2·612	160,000	16×10^{20}	16×10^{200}	16×10^{2000}
5	$\frac{1}{6}\pi^2 = 1\text{·}64493$	200	2×10^{10}	2×10^{100}	2×10^{1000}
6	1·0009846	1	11	$1\text{·}093 \times 10^{11}$	$1\text{·}093 \times 10^{111}$
7	$1 + (1\text{·}27 \times 10^{-30})$	1	1	10	$1\text{·}213 \times 10^{10}$
8	10	73	247	2214	21883
9	2	9	36	336	3325
10	10/9	3	11	101	1001
11	$e - 1 = 1\text{·}718282$	5	13	70	440
12	1·291286	3	10	57	386
13	3·234989	8	15	46	148
14	1·564468	3	6	19	58
15	1·100100	2	4	11	32
16	1·062500	2	2	3	4

* The phrase 'calculate the sum correctly to m decimal places' is used as equivalent to 'calculate with an error less than $\frac{1}{2} \times 10^{-m}$.' In the case of a very slowly convergent series the interpretation affects the numbers to a considerable extent. The numbers would be considerably more difficult to calculate were the phrase interpreted in its literal sense.

Such a series as (7) is of course exceedingly rapidly convergent *at first*, *i.e.* a very few terms suffice to give the sum correctly to a considerable number of places ; but if the sums are wanted to a very large number of places, even the series (8) proves to be far more practicable. Mr William Shanks (*Proc. Roy. Soc.*, vol. 21, p. 318) calculated the value of π to 707 places of decimals from Machin's formula

$$\pi = 16 \left(\frac{1}{5} - \frac{1}{3 \cdot 5^3} + \frac{1}{5 \cdot 5^5} - \dots \right) - 4 \left(\frac{1}{239} - \frac{1}{3 \cdot 239^3} + \dots \right).$$

He does not state the number of terms be found it necessary to use, but, in a previous calculation to 530 places, used 747 terms of the first and 219 terms of the second series. He also (*ibid.*, vol. 6, p. 397) calculated e to 205 places from the series (11).

5. Table to illustrate the divergence of the series

(1) $\dfrac{1}{\log \log 3} + \dfrac{1}{\log \log 4} + \dots$ (2) $\dfrac{1}{\log 2} + \dfrac{1}{\log 3} + \dots$

(3) $1 + \dfrac{1}{\sqrt{2}} + \dfrac{1}{\sqrt{3}} + \dots$ (4) $1 + \dfrac{1}{2} + \dfrac{1}{3} + \dots$

(5) $\dfrac{1}{2 \log 2} + \dfrac{1}{3 \log 3} + \dots$ (6) $\dfrac{1}{3 \log 3 \log \log 3} + \dfrac{1}{4 \log 4 \log \log 4}.$

Series	Number of terms required to make the sum greater than					
	3	5	10	100	1000	10^6
1	1	1	1	116	1800	$2 \cdot 6 \times 10^6$
2	3	7	20	440	7600	$1 \cdot 5 \times 10^7$
3	5	10	33	2500	$2 \cdot 5 \times 10^5$	$2 \cdot 5 \times 10^{11}$
4	11	82	12390	10^{43}	$10^{43 \times 10^3}$	$10^{43 \times 10^0}$
5	8690	$1 \cdot 3 \times 10^{29}$	10^{4300}	$10^{5 \times 10^{42}}$	——	——
6	1	60 *to* 70	$10^{10^{100}}$	——	——	——

6. Roots of certain equations.

(i) The equation $e^x = x^{1,000,000}$ has a root just larger than unity (the excess of the root over unity being practically 10^{-6}) and a large root in the neighbourhood of 16,610,800. The equation $e^x = 1,000,000 \, x^{1,000,000}$ has roots nearly equal to those of the above. The one near unity is practically $12 \cdot 82 \times 10^{-6}$ less than unity, while the large root exceeds the root of the above equation by about $13 \cdot 82$.

(ii)   The equation $e^{x^2} = x^{10^{10}}$ has a root somewhere near 357,500.

(iii)   The equation $e^{e^x} = 10^{10} x^{10} e^{10^{10} x^{10}}$ has a root near 64·7. The root differs by less than 10^{-26} from the corresponding root of $e^{e^x} = 10^{10} x^{10}$. The corresponding root of $e^x = x^{10}$ is about 35·8.

(iv)   The positive roots of $x^{x^x} = 1,000,000$ and $x^{x^x} = 10^{1,000,000}$ are approximately 2·68 and 7·11.

(v)   If $x^{10} = 10^y$, then for $x = 100$, $y = 20$; and for $x = 10^{10}$, $y = 100$. If $x^{10^{10}} = 10^{10^y}$, then for $x = 100$, $y = 10·30$; for $x = 10^{10}$, $y = 11$; and for $x = 10^{10^{10}}$, $y = 20$. If $x^{10^{10^{10}}} = 10^{10^{10^y}}$, then for $x = 10^{10}$, $y = 10 + (4·3 \times 10^{-11})$; for $x = 10^{10^{10}}$, $y = 10 + (4·3 \times 10^{-10})$; and for $x = 10^{10^{10^{10}}}$, $y = 10·30$.

7.   Some numbers of physics.

The distance to α Centauri is roughly 26,000,000,000,000 miles or $1·65 \times 10^{18}$ inches. The number of inches lies between 19! and 20! and is approximately equal to $e^{e^{3·74}}$ or 16^{e^e}. Again, writing 15 letters to the inch (an average size in print) a line to the star would be sufficient for the writing at length of $10^{2·47 \times 10^{19}}$. The latter number is approximately equal to $(14 \times 10^{17})!$, $e^{e^{e^{3·83}}}$, or $(10^{6·5 \times 10^{12}})^{e^e}$.

If we take the distance to the end of the visible universe to be that through which light travels in 10,000 years, we find that this distance when expressed in wave-lengths of sodium light is measured roughly by the numbers

$$1·6 \times 10^{26}, \quad 26!, \quad e^{e^{4·10}}, \quad 3·29^{3·29^{3·29}}.$$

If we assume the average distance between the centres of two adjacent molecules of the earth's substance to be 10^{-8} cm., we find that the number of molecules in the earth is roughly

$$10·8 \times 10^{50}, \quad 42!, \quad e^{e^{4·77}}, \quad 3·56^{3·56^{3·56}}.$$

CAMBRIDGE : PRINTED BY JOHN CLAY, M.A. AT THE UNIVERSITY PRESS